INTERNATIONAL
WILDLIFE
ENCYCLOPEDIA

THIRD EDITION

Volume 16

RIF–SEA

Marshall Cavendish Corporation
99 White Plains Road
Tarrytown, New York 10591–9001

Website: www.marshallcavendish.com

Library of Congress Cataloging-in-Publication Data

Burton, Maurice, 1898-
    International wildlife encyclopedia / [Maurice Burton, Robert Burton] .-- 3rd ed.
        p. cm.
    Includes bibliographical references (p.).
    Contents: v. 1. Aardvark - barnacle goose -- v. 2. Barn owl - brow-antlered deer -- v. 3. Brown bear - cheetah -- v. 4. Chickaree - crabs -- v. 5. Crab spider - ducks and geese -- v. 6. Dugong - flounder -- v. 7. Flowerpecker - golden mole -- v. 8. Golden oriole - hartebeest -- v. 9. Harvesting ant - jackal -- v. 10. Jackdaw - lemur -- v. 11. Leopard - marten -- v. 12. Martial eagle - needlefish -- v. 13. Newt - paradise fish -- v. 14. Paradoxical frog - poorwill -- v. 15. Porbeagle - rice rat -- v. 16. Rifleman - sea slug -- v. 17. Sea snake - sole -- v. 18. Solenodon - swan -- v. 19. Sweetfish - tree snake -- v. 20. Tree squirrel - water spider -- v. 21. Water vole - zorille -- v. 22. Index volume.
    ISBN 0-7614-7266-5 (set) -- ISBN 0-7614-7267-3 (v. 1) -- ISBN 0-7614-7268-1 (v. 2) -- ISBN 0-7614-7269-X (v. 3) -- ISBN 0-7614-7270-3 (v. 4) -- ISBN 0-7614-7271-1 (v. 5) -- ISBN 0-7614-7272-X (v. 6) -- ISBN 0-7614-7273-8 (v. 7) -- ISBN 0-7614-7274-6 (v. 8) -- ISBN 0-7614-7275-4 (v. 9) -- ISBN 0-7614-7276-2 (v. 10) -- ISBN 0-7614-7277-0 (v. 11) -- ISBN 0-7614-7278-9 (v. 12) -- ISBN 0-7614-7279-7 (v. 13) -- ISBN 0-7614-7280-0 (v. 14) -- ISBN 0-7614-7281-9 (v. 15) -- ISBN 0-7614-7282-7 (v. 16) -- ISBN 0-7614-7283-5 (v. 17) -- ISBN 0-7614-7284-3 (v. 18) -- ISBN 0-7614-7285-1 (v. 19) -- ISBN 0-7614-7286-X (v. 20) -- ISBN 0-7614-7287-8 (v. 21) -- ISBN 0-7614-7288-6 (v. 22)
    1. Zoology -- Dictionaries.   I. Burton, Robert, 1941- . II. Title.

QL9 .B796 2002
590'.3--dc21

                                                          2001017458

Printed in Malaysia
Bound in the United States of America

07 06 05 04 03 02 01   8 7 6 5 4 3 2 1

Brown Partworks
Project editor: Ben Hoare
Associate editors: Lesley Campbell-Wright, Rob Dimery, Robert Houston, Jane Lanigan, Sally McFall, Chris Marshall, Paul Thompson, Matthew D. S. Turner
Managing editor: Tim Cooke
Designer: Paul Griffin
Picture researchers: Brenda Clynch, Becky Cox
Illustrators: Ian Lycett, Catherine Ward
Indexer: Kay Ollerenshaw

Marshall Cavendish Corporation
Editorial director: Paul Bernabeo

Authors and Consultants

Dr. Roger Avery, BSc, PhD (University of Bristol)

Rob Cave, BA (University of Plymouth)

Fergus Collins, BA (University of Liverpool)

Dr. Julia J. Day, BSc (University of Bristol), PhD (University of London)

Tom Day, BA, MA (University of Cambridge), MSc (University of Southampton)

Bridget Giles, BA (University of London)

Leon Gray, BSc (University of London)

Tim Harris, BSc (University of Reading)

Richard Hoey, BSc, MPhil (University of Manchester), MSc (University of London)

Dr. Terry J. Holt, BSc, PhD (University of Liverpool)

Dr. Robert D. Houston, BA, MA (University of Oxford), PhD (University of Bristol)

Steve Hurley, BSc (University of London), MRes (University of York)

Tom Jackson, BSc (University of Bristol)

E. Vicky Jenkins, BSc (University of Edinburgh), MSc (University of Aberdeen)

Dr. Jamie McDonald, BSc (University of York), PhD (University of Birmingham)

Dr. Robbie A. McDonald, BSc (University of St. Andrews), PhD (University of Bristol)

Dr. James W. R. Martin, BSc (University of Leeds), PhD (University of Bristol)

Dr. Tabetha Newman, BSc, PhD (University of Bristol)

Dr. J. Pimenta, BSc (University of London), PhD (University of Bristol)

Dr. Kieren Pitts, BSc, MSc (University of Exeter), PhD (University of Bristol)

Dr. Stephen J. Rossiter, BSc (University of Sussex), PhD (University of Bristol)

Dr. Sugoto Roy, PhD (University of Bristol)

Dr. Adrian Seymour, BSc, PhD (University of Bristol)

Dr. Salma H. A. Shalla, BSc, MSc, PhD (Suez Canal University, Egypt)

Dr. S. Stefanni, PhD (University of Bristol)

Steve Swaby, BA (University of Exeter)

Matthew D. S. Turner, BA (University of Loughborough), FZSL (Fellow of the Zoological Society of London)

Alastair Ward, BSc (University of Glasgow), MRes (University of York)

Dr. Michael J. Weedon, BSc, MSc, PhD (University of Bristol)

Alwyne Wheeler, former Head of the Fish Section, Natural History Museum, London

Contents

RIFLEMAN

THE RIFLEMAN IS THE BEST known of the small family of New Zealand wrens. They are not related to the true wrens but are thought by some to be related to the pittas. The rifleman is tiny, under 4 inches (10 cm) long with a short tail and needlelike bill. The back, neck and crown of the male are bright green. The rump is yellow, the tail is black and the flight feathers are bluish green. In both sexes, there is a white streak over the eyes (the supercilium, or eyebrow), and the chin is white.

The only other member of the New Zealand wren family is the rock wren, or South Island wren, *Xenicus gilviventris*. It is similar in appearance to the rifleman. Riflemen are fairly common on both North and South Islands of New Zealand and on adjacent islands. The rock wren is restricted to South Island.

The typical feeding habitat of the rifleman is the moss-covered surface of a tree. When it finds an insect, it can be seen arching its back to extract the prey from its hiding place in the moss.

The bush wren, or South Island bush wren, *X. longipes*, was last seen in 1972, on South Island, and is now thought to be extinct. It was once abundant on both islands, but it declined rapidly following the introduction of predatory mammals such as cats and stoats. The subspecies of bush wren found on North Island was last seen in 1949. The sad case of the South Island bush wren is the fifth extinction of New Zealand wrens in recent times.

Adapting to change

The rifleman lives in open forests, particularly in the mountain beech forests, but often comes into open country or bush. It is now found in suburban parks and other modified habitats with remnants of native forest. Its presence is given away by its high-pitched *zee* as it flies from tree to tree. The relative abundance of the rifleman appears to be due to its ability to adapt to the changing countryside. The rock and bush wrens have not shown the same adaptability.

Tree creeping

The rifleman and its relatives are insect-eaters and have the fine, pointed bills typical of birds such as the European robin, *Erithacus rubecula*, that hunt insects among plants. The feeding behavior of riflemen is very much like that of treecreepers, *Certhia familiaris*. Starting at the foot of a tree, a rifleman hops up the trunk in spirals, searching for insects and spiders in the bark or among mosses and lichen. When it has ascended 20–30 feet (6–9 m), it glides to the base of the next tree. Riflemen also turn over dead leaves on the ground in search of food, and they sometimes mix with flocks of white-eyes, *Zosterops lateralis*, which search among branches and foliage for insects. Rock wrens feed among the alpine vegetation of screes and moraines, and eat insects, spiders and the fruits of alpine plants.

Loose nest

The rifleman builds its nest in a hole or crevice, perhaps under bark, near the ground, making a tiny entrance. Within the cavity is a loosely woven globe of fine roots, grasses and skeleton leaves, lined with feathers or fur. The nest may be 6 inches (15 cm) high and 4 inches

RIFLEMAN

CLASS	**Aves**
ORDER	**Passeriformes**
FAMILY	**Acanthasittidae**
GENUS AND SPECIES	***Acanthisitta chloris***

ALTERNATIVE NAMES
Titipounamu (Maori); thumbie

WEIGHT
⅙–⅕ oz. (5–6 g)

LENGTH
Head to tail: 3½ in. (8.5 cm)

DISTINCTIVE FEATURES
Tiny size; very thin bill; white eyebrow and chin; short tail. Male: green nape, crown and upperparts; bluish green flight feathers; reddish tinge to forehead; yellow rump; black tail. Female: brown upperparts.

DIET
Insects taken from bark, and from mosses and lichens on trees

BREEDING
Age at first breeding: 1 year; breeding season: eggs laid August–January; number of eggs: 3 or 4; incubation period: 20 days; fledging period: 24 days; breeding interval: 1 or 2 broods per year

LIFE SPAN
Not known

HABITAT
Open forest, scrubland, suburban parks, hedgerows and plantations

DISTRIBUTION
New Zealand only: southern two-thirds of North Island; through most of South Island and Stewart Island; several other, smaller offshore islands

STATUS
Common

Rifleman

(10 cm) wide, with a chamber only 2½ inches (6.5 cm) across. Both parents incubate and feed the chicks. The fledglings stay with the parents for some time, but two broods may be raised in one season. The rock wren has similar breeding habits to those of the rifleman, but the rock wren nests among boulders.

Unlike the other, more unfortunate species of New Zealand wrens, the rifleman has adapted successfully to a range of modified habitats, including parks.

One predator was enough

The increasing rarity of New Zealand wrens is partly due to the clearing of woodland and partly due to introduced predators. They are especially vulnerable because they live near the ground and are not strong fliers. The fourth New Zealand wren, *Xenicus lyalli*, became extinct before the beginning of the 20th century. It was confined to Stephen Island, a small island 2 miles (3.2 km) off South Island, when it was discovered by a lighthouse keeper whose cat brought in 11 of them. These specimens were sent to ornithologists in Europe. No more were found, and it seems the cat had wiped out an entire species. The only record of the bird's habits is a statement from the lighthouse keeper to the effect that it appeared in the evening and was flightless. If this is true, the Stephen Island wren was the only flightless passerine, or perching, bird. It also had the smallest range of any known bird, as Stephen Island is only 1 mile (1.6 km) long.

RIGHT WHALE

Right whales are known for their habit of raising their flukes above the water to bring them crashing down on the surface, a technique known as lobtailing.

RIGHT WHALES WERE THE first whales to be hunted in large numbers and were virtually wiped out before hunting of them was abandoned. Their common name derives from the fact that in the early days of whaling they were the most suitable kind of whales to catch, as they swam slowly, floated after they were killed and yielded a large amount of oil and baleen (a horny substance found in two rows of plates attached to the upper jaws of some whales).

One hundred and fifty years ago the bowhead whale, *Balaena mysticetus*, was widely known as the common whale, because it was so numerous. Now it is one of the rarest whale species. None of the four species of right whales is common. They are all baleen whales, differing from the larger rorquals in lacking the grooves in the throat and in having no dorsal fin, except for the pygmy right whale, *Caperea marginata*. Their heads are huge and may be one-third of the total body length. The size of the head and the lower lips rising on either side of the narrow, arched upper jaws gives a striking appearance. When the mouth is opened, the lower lips form the sides of a tunnel. The baleen plates are extremely long, being up to 15 feet (4.5 m), although they usually are 10–12 feet (3–3.6 m) in the bowhead

whale, which has the most strongly arched jaws. There may be 350 plates on just one side of the upper jaw. When the mouth is closed, the plates fold on the floor of the mouth.

The bowhead once lived in the Arctic seas around North America, Europe and Asia, but the remnants of the population are now confined to the Canadian Arctic. It measures 50–60 feet (15–18 m) and is black with a cream chin and sometimes has white underparts. The lower jaw is U-shaped and the upper jaw is strongly arched. The northern right whale, *Eubalaena glacialis*, is found in the North Atlantic and North Pacific Oceans. The southern right whale, *E. australis*, is found across the Southern, South Pacific and South Atlantic Oceans, but not in tropical waters. They grow up to 54 feet (16 m) and are black, sometimes with a white belly.

The upper jaw of the right whales is less arched than that of the bowhead whale, and on the front of the head is the bonnet, a horny growth often infested with parasitic worms and crustaceans. The pygmy right whale is confined to the southern seas. It measures only 20 feet (6 m), is dark blue or gray with lighter underparts, and has a small dorsal fin. The inside of its mouth is pure white.

RIGHT WHALES

CLASS **Mammalia**

ORDER **Cetacea**

FAMILY **Balaenidae**

GENUS AND SPECIES **Bowhead whale, *Balaena mysticetus*; northern right whale, *Eubalaena glacialis*; southern right whale, *E. australis*; pygmy right whale, *Caperea marginata***

ALTERNATIVE NAMES
Arctic whale, Greenland whale (*B. mysticetus*); black right whale (*E. glacialis*, *E. australis*)

WEIGHT
3⅓–110 tons (3–100 tonnes)

LENGTH
Head and body: 18–60 ft. (5.5–18 m)

DISTINCTIVE FEATURES
Very arched mouthline; large, rotund body (*B. mysticetus* and *E. glacialis* only); black skin with some ventral white markings (*E. glacialis* and *E. australis* only); *C. marginata* is only species with dorsal fin

DIET
Mainly krill; also other crustaceans and invertebrates

BREEDING
***E. australis.* Age at first breeding: about 10 years; breeding season: peaks in August and October; number of young: 1; gestation period: about 360 days; breeding interval: 2–5 years.**

LIFE SPAN
Up to about 100 years

HABITAT
Open oceans and coastal waters

DISTRIBUTION
***E. glacialis*: northern Atlantic and Pacific Oceans. *E. australis*: southern oceans.**

STATUS
***B. mysticetus* and *E. glacialis*: endangered; *E. australis*: conservation dependent**

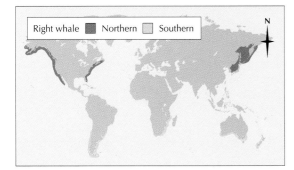

Right whale ■ Northern □ Southern

The southern right whale (above) was nearly extinct by the start of the 20th century. Protected by law since 1935, it remains one of the most endangered of all the world's cetaceans.

Slow swimmers

At one time right and bowhead whales lived in large herds, sometimes consisting of 100 to 200 individuals, but now they are seen only singly or in small groups. The pygmy right whale seems never to have been common. Migrations are limited; those of the bowhead whale seem to be determined by the seasonal movements of pack ice in the Arctic Sea. Right whales in the Pacific move from Japan to the Bering Sea in the summer; there is a parallel movement in the Atlantic.

Right whales swim very slowly, cruising at about 5 miles per hour (8 km/h), although they are capable of bursts of 10 miles per hour (16 km/h) when hard pressed. They can submerge for up to an hour, but usually dive for 10–15 minutes and then swim at the surface for 5–10 minutes, blowing once every minute. The spout is V-shaped, with the nostrils diverging more than in other whales.

Continuous feeding

The main food of right whales is krill (or the crustaceans *Thysanoessa inermis* and *Meganyctiphanes norvegica*, the equivalent of krill in northern seas), together with copepods and planktonic mollusks. Although this food is strained from the water by the baleen, the method used is different from that used by rorquals such as the blue whale, *Balaenoptera musculus* (discussed elsewhere in this encyclopedia). The rorquals have short baleen plates, and their mouths can be distended by dropping the floor. They feed by taking great gulps of water and squirting it through the baleen. Right whales, however, have very long baleen plates, and feed by swimming with open mouths, the water pouring out through the baleen, which catches the krill and other small animals. Every so often the mouth is closed and the food is swept back to the throat. The pygmy whale is thought to feed near the bottom of the sea in contrast to other right whales, which all feed at or just below the surface of the sea.

There is some scientific debate about the breeding habits of right whales. In the North Atlantic the bowhead whale seems to mate in February and March, but mating has also been recorded at the end of the summer. A single calf is born 13–14 months later and suckled for 1 year.

Whaling traditions

The first people to hunt whales systematically were the Basques living on the shores of the Bay of Biscay, which lies off France and Spain. The Biscayan right whale was abundant at the time and was easy to catch because it moved so slowly. The faster whales, such as rorquals, could only be hunted when bigger and faster boats were introduced. The Basques started to hunt right whales in the 11th century, capturing them in the winter months when they came close to the shore. The right whales of the Bay of Biscay were gradually wiped out, and whaling subsequently spread to Portugal and England, where the whale was regarded as a royal fish and the head of each one caught was the king's property.

About the time that Europeans were first exploring across the Atlantic, the Basques reached Newfoundland to hunt the bowhead whale. They were followed by the Dutch and English, who at first employed Basques to catch the whales. Whaling grounds off the coasts of Iceland, northern Norway, the Davis Strait, Spitsbergen and other places were discovered, exploited and left devastated. The final decline occurred in the mid-19th century. The last stronghold of the bowhead whale in the Bering Sea was fished out by the United States in the 19th century.

Right whales are now fully protected by law. However, their populations remain small, and some species, such as the northern right whale and the bowhead, are endangered. The Brazilian government has recently agreed to the establishment of the country's first whale sanctuary. It is located off the southern coast of Santa Catarina and will be a haven for the southern right whale.

Right whales are notable for the wartlike bumps, or callosities, that cover the head. Each individual has a unique pattern of bumps.

RINGED SEAL

ONE OF THE MOST COMMON seals is the ringed seal, *Pusa hispida*. Although confined almost entirely to Arctic seas, its population has been estimated at 6 to 7 million. This small seal measures about 5 feet (1.5 m) from nose to tail and weighs about 200 pounds (91 kg). The color of the fur is very variable but is usually gray with dark spots ringed with white; hence its name.

Ringed seals are found all around the Arctic coasts and are often seen off the northern coast of Iceland in winter, stragglers reaching the British Isles and even Finisterre. An isolated population lives in the Baltic Sea.

The Caspian and Baikal seals, confined to the Caspian Sea and Lake Baikal respectively, are very closely related to the ringed seal. Caspian seals, *P. caspica*, which are estimated to number about 500,000, are a little smaller than ringed seals. They are grayish with dark spots sometimes ringed with white. Baikal seals, *P. sibirica*, are the smallest seals, about 4½ feet (1.4 m) long and weighing about 140 pounds (63.5 kg). They are dark brownish gray, and their numbers are estimated at 60,000 to 70,000.

Cool water preferred

Ringed seals are very much at home on the ice, preferring fast ice to loose floes, and are even found in the middle of the Arctic Sea. Caspian seals move north and south seeking cooler water. In winter they are found at the northern end of the Caspian Sea where the water is frozen, and the pups are born on the ice. As the year progresses, they move southward. Baikal seals are most common at the northern end of the lake. They spend the winter in the water, breathing at air holes, but lie on the ice in the fall and spring.

Fish and crustaceans

Although small fish such as the polar cod are eaten, the main food of ringed seals is planktonic crustaceans such as amphipods and opossum shrimps. Ringed seals do not have the same cusped teeth as the crabeater seal, *Lobodon carcinophagus* (discussed elsewhere), for sieving planktonic animals. It seems likely that they pick out large amphipods, up to 2½ inches (6.5 cm) long, individually. Since they also eat small animals in quantity, however, it is quite probable that ringed seals strain the seawater between their teeth to catch them. Caspian seals and Baikal seals have similar diets, eating sprats, gobies and herrings, which are commonly found in these enormous inland lakes.

Born under snow

The pups are born in spring, from mid-March to mid-April. They are always born on ice that is firmly anchored to land rather than on drifting pack ice. The mother makes a den under the snow or uses a natural cavity. There is an air hole in the floor of this cavity, through which she can slip into the water to feed. Within the den the mother and pup are protected from bad weather and are relatively safe from predators, although polar bears and foxes sometimes dig into the den and kill the pups.

At birth the pup weighs about 10 pounds (4.5 kg) and is 2 feet (60 cm) long. Its first coat of fur is long and white, but this is shed after two weeks and is replaced by a dark coat. The pup is suckled for nearly 2 months, during which time its mother will have mated again.

The Baikal seal bears its pups in February or March, the females having spent the winter in dens under the snow instead of in the water like

Ringed seal pups weighing about 10 pounds (4.5 kg) are always born on ice that is attached to land, never on drifting floes.

RINGED SEAL AND RELATIVES

CLASS	**Mammalia**
ORDER	**Pinnipedia**
FAMILY	**Phocidae**

GENUS AND SPECIES   **Ringed seal, *Pusa hispida* (details below); Caspian seal, *P. caspica*; Baikal seal, *P. sibirica***

WEIGHT
110–253 lb. (50–115 kg)

LENGTH
Head and body: 3⅗–5⅗ ft. (1.1–1.7 m)

DISTINCTIVE FEATURES
Distinctive black spots, often ringed with a lighter color; variable coat color, usually gray

DIET
Fish; crustaceans and other invertebrates

BREEDING
Age at first breeding: 5–7 years; breeding season: April–May; gestation period: 11 months, including 3½ months of delayed implantation; number of young: usually 1; breeding interval: 1–2 years

LIFE SPAN
Up to 45 years

HABITAT
Polar fast ice and open waters

DISTRIBUTION
Around Arctic coasts

STATUS
Common, although some subspecies endangered or vulnerable

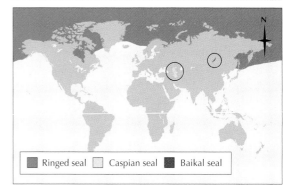

☐ Ringed seal   ☐ Caspian seal   ■ Baikal seal

The Caspian seal is confined to the Caspian Sea, a vast inland lake. There are several theories as to how the seal arrived in a landlocked water.

the males and juveniles. The Caspian seal has its pups in January and February. In every other respect the breeding of the Caspian and Baikal seals follows the pattern of the ringed seal.

Ringed by predators

Polar bears, walruses, Arctic foxes and killer whales are the main predators of ringed seals, the pups being especially vulnerable. Hunting ringed seals is an important part of the lives of many Inuits. In summer they shoot them in the water, while in winter they harpoon them at their breathing holes or stalk them as they lie on the ice. The Inuits use the seal skins for fur or leather, and eat the meat and blubber. Caspian and Baikal seals are also hunted; their white-coated pups were once especially prized, but it is now illegal to cull pups under 2 weeks old.

How did they get there?

The presence of seals in the landlocked waters of the Caspian Sea and Lake Baikal has aroused considerable speculation. They are closely related to the ringed seal of the Arctic Sea, yet there is no obvious route they could have taken to get where they are today. One theory is that they could have spread to these lakes in prehistoric times when Central Asia was covered by the Tethys Sea. Another idea is that during the Ice Age the glaciers dammed up the rivers, forming huge lakes up which the seals traveled. A study of fossil seals, however, suggests that these two seals appeared too late for their distribution to

have been influenced by the Tethys Sea and too early for the Ice Age. There may have been other unknown connections with the sea that have since disappeared, and it is possible that the seals could have reached the Caspian and Lake Baikal by swimming up rivers and even crossing land.

RINGTAIL

RELATIVE OF THE RACCOON, the ringtail, *Bassariscus astutus*, is slender and sleek and weighs about 2 pounds (900 g). It is up to almost 3 feet (90 cm) in length, of which about half is bushy tail, ringed black and white. The coat is grayish buff, darker along the back and white on the underparts. The face is foxlike, the eyes are large and white-ringed with contrasting black patches, and the ears are large and pricked. The feet are well furred, and the claws can be partly withdrawn. The ringtail ranges from southwestern Oregon and eastern Kansas to Baja California and eastern Mexico.

Among the ringtail's many alternative names is cacomistle, which derives from an original Mexican name, cacomixtle, meaning rush cat. A further Mexican name is tepemixtle, or bush cat. In the United States it has been given a variety of names, including coon cat, raccoon fox, band-tailed cat, cat-squirrel, miner's cat, mountain cat and ring-tailed cat. Its fur is valuable and is marketed as civet cat and California mink.

Related to the common ringtail is the Central American ringtail, or guayonoche, *B. sumichrasti*, about which fairly little is known. A little larger than the ringtail and more arboreal (tree-dwelling) in habit, it lives in the forests of southern Mexico and Central America as far south as western Panama. Its ears are less rounded than those of the ringtail and it has nonretractile claws.

Elusive ringtail

Nocturnal and secretive, the ringtail usually manages to keep out of sight. It sleeps by day in a den among rocks, in holes in trees or at the base of trees between buttress roots. Agile in moving among trees, it uses its tail as an aid to balance and also curls it over its back as do squirrels. The ringtail does this especially when alarmed, at the same time giving a squirrel-like scolding or barking sound.

The ringtail's food is small rodents and birds, insects and vegetable matter. It is well known around the fruit farms of California for its habit of eating the fallen fruits. At times a ringtail may come to live in a house, cottage or cabin and keep the place free of mice and rats. Prospectors used the ringtail to keep down rodent numbers in mines, from which the animal got its alternative name of miner's cat, mentioned earlier. Some people, however, consider the ringtail a pest because it is thought to kill domestic chickens.

The ringtail prefers to live in rocky places. It is an excellent climber and sheer walls are no obstacle.

RINGTAILS

CLASS	**Mammalia**
ORDER	**Carnivora**
FAMILY	**Procyonidae**

GENUS AND SPECIES **Ringtail, *Bassariscus astutus* (details given below); Central American ringtail, *B. sumichrasti***

ALTERNATIVE NAMES
Band-tailed cat; cacomistle; cat-squirrel, coon cat; miner's cat; mountain cat; raccoon fox; ring-tailed cat; tepemixtle

WEIGHT
About 2 lb. (900 g)

LENGTH
Head and body: 12–16½ in. (30–42 cm); shoulder height: 6⅓ in. (16 cm); tail: 12¼–17⅓ in. (31–44 cm)

DISTINCTIVE FEATURES
Size of domestic cat; foxlike form, with very long, furry, ringed tail

DIET
Insects, rodents, birds and vegetable matter

BREEDING
Age at first breeding: 10 months; breeding season: February–May; number of young: 1 to 5; gestation period: 51–54 days; breeding interval: probably 1 litter per year

LIFE SPAN
Up to 14 years in captivity

HABITAT
Varied, but mainly in rocky areas

DISTRIBUTION
Southwestern Oregon and eastern Kansas south to Baja California and southern Mexico

STATUS
Locally common

Although rocky places are preferred, ringtails sometimes live in woodlands, in hollow trees. Ringtails ambush their prey, pinning it down with their paws before killing it with a bite to the neck.

The gestation period in ringtails is 51–54 days. The young are born in May or June, in a moss-lined nest. The three or four babies in a litter weigh 1 ounce (28 g) each. They are born blind, with ears closed and the body covered with a downy fur. Some solid food may be taken at 3 weeks, and the eyes open at 4 weeks. The young are taken hunting at 2 months and are weaned at 4 months. Ringtails have been known to live just over 14 years in captivity.

Night-living sunbathers

Animals active by day are termed diurnal, those that come out at night, nocturnal. Even nocturnal animals, however, sometimes come out during the day, especially to sunbathe. The ringtail sunbathes, but it is seldom seen doing so. It does its basking in the tops of trees, crouched along a branch with its ringed tail dangling over the side, the rest of its body camouflaged against the bark.

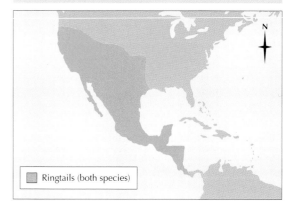

Ringtails (both species)

RIVER AND STREAM

ALTHOUGH ACCOUNTING FOR LESS than one percent of the world's surface water, the freshwater environments of rivers and streams occur in almost all areas of the world. The term river describes a flow of surface water restricted to a channel. Streams are smaller, narrower flows, often comprising the upper stretches of a river system.

While lakes and ponds contain water that is still or that has a weak current, rivers and streams contain water with a strong current or flow. This might seem obvious, but it is important because it controls which aquatic animals and plants can live there. In particular, wherever there is a continuous flow of water downstream, the growth of plankton, the crucial primary producers of lakes, is prohibited. Plankton, at the mercy of water currents, would always be carried downstream, away from their favored habitat, and would find it difficult to migrate back upstream. Plankton are absent from rivers, and only plants that manage to anchor themselves firmly to the river bed can survive.

There is a wide variation in the amount of flow in rivers and streams, from place to place and from season to season. Even in a river's energetic upper reaches, the water may idle in pools and riffles, particularly below waterfalls, for example, and here the composition of the living community can resemble that of still water in a lake or pond.

Rivers are not closed systems, and they interact with the terrestrial ecosystems through which they flow. Nutrients and organic matter are carried in by tributaries, and vegetation on the banks, both aquatic and terrestrial, can play a part in the river's ecosystem.

A river changes greatly between its source and its mouth, and offers a rich range of habitats for a diverse range of species along the way.

The source of streams

Most rivers have rather inauspicious beginnings. Their source, a spring, is typically little more than a damp patch located high in the mountains. Water comes to the surface as springs for a number of reasons. An underground flow of water might reach some impermeable rock that forces it to the surface. Near the spring, a

The upper course of the Columbia River in Oregon is little more than a cascading stream in which only aquatic organisms that can combat the strong current can survive. Mosses and liverworts maintain a hold on the banks.

For the pumpkinseed, Ledomis gibbosus, mating and spawning are triggered by rising temperatures. This ensures there is enough food available by the time the young hatch. The female lays a clutch of eggs into a hollow dug by the male.

stream is usually cold, constant in temperature and fast flowing, ideal conditions for oxygenation of the water. Since this environment changes little, we find a specialized living community here, often composed of organisms that have likewise changed little for hundreds of millions of years. Mayflies and stoneflies are examples, said to be some of the most primitive insects. Stoneflies, the carnivorous larvae of which dwell in the waters, are restricted to the cleanest, least polluted streams and rivers.

River systems

The shape and development, or morphometry, of a river system is a feature that geologists use to classify it. Devised by A. N. Strahler, the main system of classification identifies fingertip tributaries from underground spas and similar sources as first order streams. Where two of these first order streams meet, a second order stream is created, and so on until the river's highest order stream reaches the sea. These drainage patterns can take many forms but the most frequent is the treelike dendric pattern. The amount of water a river carries depends on many factors; perhaps the most important of these is the composition of the rock and soil that make

up the surrounding landscapes. Impervious rock will typically deliver roughly 75 percent of rainfall to a surface river system, while porous rocks and ground soil absorb far more water, holding back up to 85 percent of rainwater from immediate discharge into the river.

Babbling brooks

Rivers change continuously along the route to their mouths, changes often categorized into three stages: the upper course of shallow, fast-moving first order tributary streams, the middle course of second and third order streams where the river widens and becomes more stable, and finally the lower course were the river begins to meander and curve toward the sea.

At the start of the journey, in the upper course, the streams flow with a lot of energy. The slopes of their courses are steep, and the water current has so much energy that all but the largest stones are swept along as sediment. The streams cause much more erosion than deposition. Animals and plants living in the strong flow of water in the upper course live with the constant risk of getting carried away by the current. If animals and plants can solve this problem by anchoring firmly, swimming strongly or by hiding in sheltered crevices, they can reap the benefits. A strong current brings a constant supply of food and mineral salts, enabling some animals, for instance blackfly larvae (family Simuliidae), to be filter feeders. It also offers an ideal environment for creatures such as the brown trout, *Salmo trutta*, that prefer turbulent, well-oxygenated water.

In the upper courses, the main plants gaining a foothold on the rocky stream bed are mosses and liverworts (class Musci and class Hepaticae). These provide an ample home and sustenance to many tiny protozoa and rotifers. Insects such as the caddis flies and crane flies (order Trichoptera and family Tipulidae) not only thrive in this environment, but have also adapted to fill a variety of niches throughout the river's course. For example, while many species of caddis flies survive here as plant feeders, members of the genus *Rhyacophila* are free-living, tough-bodied predators, preying on other insects.

Mature rivers

As the river reaches flatter country the current slackens, allowing temperatures to fluctuate, and the volume of water increases. Many of the creatures from the upper course can survive here, with insects again thriving. Caddis flies continue to serve as an important link in the food chain, with many species of fish and birds feeding on the adults and larvae. To combat this the vegetarian species of caddis flies build protective cases out of bits of twigs, leaves and gravel to keep out many unwelcome predators, while their larvae are most often found hiding in the bed of fast-moving streams.

By the lower course the river has widened and deepened. With its flow reduced even further it is forced to deposit some of the heavier sediment it is carrying and begins to curve around to form large bends or meanders. Plants can establish themselves here, and the riverbanks offer an environment that is similar to the littoral zone of ponds. While the fluctuating oxygen level prevents many fish in the upper courses from surviving, a wide variety of species still jostle for space. Fish such as the European perch, *Perca fluviatilis*, haunt the shadows alongside fresh-water crayfish such as the white-clawed species *Austropotamobius pallipes*, while as ever, insects hover nearby. The magnificent belted kingfisher, *Megaceryle alcyon*, takes advantage of the slow-moving lower reaches and me-anders of rivers to hunt its favored

prey, small fish and fry. Aside from these, this brightly colored, eye-catching bird will also eat caddis flies and even small crayfish. The belted kingfisher is an expert angler and can often be spotted on a branch over the water, watching intently for fish. The instant a fish is spotted the bird dives headlong into the water, grabbing its prey with its open, daggerlike bill. The action takes seconds.

Meanders and oxbows

Meanders are caused by irregularities in the riverbed that cause a deflection within the stream's flow, eroding a river's bank into a bend. Water flowing at the outside of such a bend flows a little quicker, further eroding the bank as it picks up sediment. On the inside of the bend the water moves more slowly and tends to deposit its sediment. Over time, these processes create a migrating meander, its curve becoming more and more exaggerated. Eventually a meander may become so eroded that the river will cut off the meander's neck. The faster-flowing, straighter area now becomes the main route for the river, with sediment slowly accumulating around the ends of the now redundant meander, silting it up and leaving a curved oxbow lake. Isolated from the constant flow of the water, species adapted to running water may die, while still-water species colonize. In this way, the living community of the oxbow lake alters to a more pond-adapted assemblage.

After a kill, the belted kingfisher (female, below) manipulates its catch in its bill, getting it into the right position for swallowing.

The Amazon River drains the largest river basin in the world, roughly equal to half of South America. Rising in the Andes just 100 miles (160 km) from the Pacific, the river's estuary is over 4,000 miles (6,400 km) away in the Atlantic. It is home to the greatest diversity of life found in any river.

Deltas

When a river enters the sea its velocity rapidly decreases and it deposits the last of its sediment. If conditions are favorable, this sediment will build up to form a delta, a rich, fertile landform at the river's mouth that is ideal for farming. The delta of the Nile River has been used for farming for millennia, although since the building of the Aswan High Dam in Egypt much of the Nile's sediment has been prevented from reaching the sea, causing the fertile delta to shrink.

Not all river mouths are favorable for delta formation. Despite an estimated discharge of 1.65 million tons (1.5 million tonnes) of sediment every day, the Amazon has not developed a delta because of the force of the 1 billion tons (900 million tonnes) of water it discharges each year. Instead, the sediment is dispersed into a layer of ridged dunes on the estuary bed, 1¼ miles (2 km) thick.

It is not inevitable that a river system will reach the sea. The Okavango Delta in south-central Africa is a network of distributaries and wetlands that fan out into the Kalahari Desert, while the river goes no farther. Likewise, many rivers end in Lake Chad in Central Africa, but their waters evaporate, never reaching the sea.

Floods

River flow is usually seasonal, sometimes drastically so. At peak river flow, for instance in the rainy season, the channel of some rivers is too small, and the water escapes to inundate the floodplain, the area of relatively flat land that borders a river. In contrast, the lowest seasonal flow may approach or equal zero, the rivers drying out until rainfall increases or snows melt.

Activities of humans can further increase these extremes of river flow. Drainage of river basins has been speeded by urbanization and deforestation. The soil and vegetation absorb less water, allowing greater amounts to flow straight into the river and making it more liable to flood. On the other hand, in some rivers, meanders are removed by humans in an effort to regulate river flow. The Mississippi River has had 16 meanders deliberately cut off in the 370-mile (600-km) stretch between Memphis and Baton Rouge. This has allowed water to pass through the river more quickly, reducing the risk of flooding.

Almost all rivers have been affected by human action; their courses have been diverted for irrigation or dammed for power generation, channels have been dredged for navigation, their waters have been polluted with toxic chemicals and other waste products. With so many changes affecting rivers it is difficult to predict the effect of human activities on the risk of flooding and how river environments will react.

Estuaries

When a river finally reaches the sea, the most profound change takes place. The flowing fresh water of the river mixes with the saline (or salty) water of the sea. Only a few of the species that inhabit the freshwater stretches of the river can survive. One of these is the Atlantic salmon, *Salmo salar*, which lives its adult life in the sea, returning to fresh water to spawn. Other species such as the freshwater-dwelling Amazon river dolphin, *Inia geoffrensis*, are thought to have moved away from the sea and adapted to their new environment from a saltwater species that penetrated far upstream.

RIVER DOLPHIN

MARINE DOLPHINS AND porpoises sometimes swim far up rivers. However, four dolphin species live only in rivers. These are the boto or Amazon river dolphin (*Inia geoffrensis*); the baiji or Chinese river dolphin (*Lipotes vexillifer*); the Indus dolphin (*Platanista minor*); and the Ganges dolphin (*P. gangetica*). A closely related species, the franciscana, *Pontoporia blainvillei*, lives in the coastal waters off Brazil and Argentina, being particularly abundant near estuaries.

All river dolphins are small, the largest reaching a length of about 8¼ feet (2.5 m) and the smallest being only 4¼ feet (1.3 m). All have long, slender jaws that form a narrow beak, and a distinct neck. They generally are light in color, often being gray, pale brown, pink or white. Their eyes are small and weak; the Indus and Ganges dolphins are effectively blind.

Finding their way in the dark

River dolphins seem to have a similar lifestyle to marine dolphins and porpoises, though they are less active. They are either solitary or live in very small groups of three or four; sometimes they may form larger groups of up to a dozen. They swim slowly, usually at about 2 knots (3.7 km/h), though they can reach 10 knots (18.5 km/h). The Ganges dolphin usually surfaces to breathe about every 30 seconds. It may merely bring its blowhole level with the surface or it may leap up out of the water and then plunge back in again.

River dolphins live in muddy water, further reducing the effectiveness of their eyes, and have no sense of smell. Echolocation (sensing objects or prey using reflected sound waves) is the most useful means of orientation for them to employ. As with porpoises and dolphins that are known to use echolocation, it is almost impossible to catch river dolphins in nets made of mesh less than 10 inches (25 cm) wide. Most river dolphins detect small-meshed nets in time to leap over them, demonstrating that they use some sense other than sight both to avoid rocks and other objects in rivers and to catch their prey.

Lower jaw as probe

Most dolphins apparently catch their prey, mostly fish and crustaceans, by seizing it in the jaws, and the river dolphins catch some of their food this way. The boto, for example, uses this method to capture piranha. However, freshwater dolphins also frequently dig in the mud for food, using their beaklike snout to disturb mudfish and freshwater shrimps. The two halves of their lower jaws are unusual in being joined along almost the whole of their length, and this probably provides greater strength for probing mud. The teeth are slender and conical, varying in number from 52 in each jaw of the Ganges dolphin to 55 in each jaw of the franciscana dolphin. They are used for seizing the slippery prey, which is swallowed whole.

The breeding behavior of river dolphins seems to be similar to that of marine dolphins. In the Ganges dolphin breeding occurs from April to July, the young being born 8–9 months later. The young sometimes ride on their mother's back. The flippers of freshwater dolphins are broader, shorter and more handlike than those of marine dolphins and may provide the young with support in the water.

The baiji is the most threatened of all cetaceans. River damming, collisions with boats and entanglement in fishing gear may have reduced their numbers to as few as 200.

River dolphins differ from marine dolphins in having an extended, beaklike snout, a more distinct neck and tiny vestigial eyes.

RIVER DOLPHINS

CLASS	**Mammalia**
ORDER	**Cetacea**
FAMILY	**Iniidae, Pontoporiidae and Platanistidae**

GENUS AND SPECIES **Iniidae: boto or Amazon river dolphin, *Inia geoffrensis*. Pontoporiidae: baiji or Chinese river dolphin, *Lipotes vexillifer*; franciscana or La Plata dolphin, *Pontoporia blainvillei*. Platanistidae: Indus dolphin, *Platanista minor*; Ganges dolphin, *P. gangetica*.**

WEIGHT
66–352 lb. (30–160 kg)

LENGTH
Head and body: 4¼–8¼ ft. (1.3–2.5 m)

DISTINCTIVE FEATURES
Long, thin beak; distinct neck; tiny eyes; smaller and slower than oceanic dolphins

DIET
Fish and crustaceans

BREEDING
***I. geoffrensis*. Breeding season: probably October and November; number of young: 1; gestation period: 10–11 months; breeding interval: not known.**

LIFE SPAN
Up to 30 years

HABITAT
Freshwater rivers; flooded forests; coastlines

DISTRIBUTION
Northern South America; southern Asia. *P. blainvillei*: eastern coast of South America.

STATUS
Uncertain. *I. geoffrensis*: vulnerable. *L. vexillifer*: critically endangered. *P. minor*, *P. gangetica*: endangered.

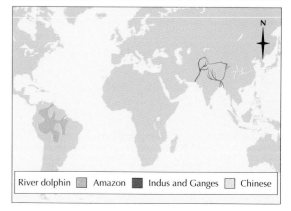

River dolphin ▢ Amazon ▨ Indus and Ganges ▢ Chinese

River dolphins are fairly free from predators, apart from humans. The franciscana is credited locally with bringing the bodies of drowned people to shore and is held sacred. However, the Ganges dolphin is killed for its flesh and blubber, the oil from the latter being valued as a liniment for rheumatism and for strengthening the muscles of the back and loins. It is also used in lamps. The boto, by contrast, is not hunted, because of a local superstition that to use its oil in lamps causes blindness. This is based on the tiny, almost useless eyes of this dolphin.

Ancestry of the river dolphins

It has often been assumed that because whales, porpoises and dolphins are so obviously descended from land animals, their ancestors must have returned to the sea shortly after emerging from it. In this context, animals such as river dolphins are something of an oddity, and in the past biologists have reasoned that the animals have come upriver from the sea. In fact, the ancestors of dolphins could just as easily have left the land for the rivers and later invaded the sea. This is supported by the primitive characteristics exhibited in the skeletons of the freshwater dolphins. In most whales, porpoises and dolphins the seven neck vertebrae are squashed together and fused, and the animals have no visible neck. In freshwater dolphins the vertebrae are separate and there are still vestiges of a neck. Also, the skull of a freshwater dolphin has not undergone the same fundamental changes as the skulls of its marine relatives, and in several ways it is more like the skull of the extinct dolphin *Squalodon*, which lived 15 million years ago.

RIVER OTTER

THE VARIOUS SPECIES OF otters are all much alike in appearance and habits. They are long-bodied, short-legged mammals, with a stout tail thickened at the root and tapering toward the tip. There is a pair of scent glands under the tail. The head is flattened, with a broad muzzle and numerous bristling whiskers. The ears are small and almost hidden in the fur. The sleek, dark brown fur consists of a close fawn underfur, which is waterproof, and an outer layer of long, stiff guard hairs, which are gray at their bases and dark gray, brown or black at their tips. The throat is whitish and the underparts are pale brown. Each foot has five toes, bearing claws in most species; the forefeet are small and the hind feet are large and webbed, except in the genus *Aonyx*.

The European or common otter, *Lutra lutra*, ranges across Europe and parts of Asia to Japan and the Kurile Islands. It is 4 feet (1.2 m) long, including the tail, but may reach 5½ feet (1.7 m), and weighs up to 25 pounds (11.3 kg). The bitch (female) is smaller than the dog (male) otter. The Canadian otter of Canada and the United States, *L. canadensis*, is very similar to the European otter but has more variation in size. It is sometimes spoken of as the river otter to distinguish it from the sea otter, *Enhydra lutris*, a markedly different animal. The small-clawed otter, *Amblonyx cinerea*, of India and Southeast Asia, is sometimes smaller than the European species. The clawless otter, *Aonyx capensis*, of western and southern Africa, is larger and a marsh dweller, feeding on frogs and mollusks. The giant Brazilian otter, *Pteronura brasiliensis*, is the largest of all the otters. It reaches 6½ feet (2 m) in length and has a tail that is flattened from side to side.

Solitary and elusive

Except during the mating season most otters are solitary, extremely elusive and secretive, and always alert for any sign of disturbance. However, giant otters and smooth-coated otters can be sociable. Otters can submerge in a flash, leaving few ripples, or they will disappear among vegetation when on land. Their ability to merge into their background on land is helped by the "boneless" contortions of the body and the changing shades of color in the coat, effected by movements and changes in the guard hairs.

River otters such as the Canadian otter, **Lutra canadensis,** *feed after dusk and before dawn. Their long whiskers help them to locate prey in poor underwater visibility.*

Otters exhibit playful behavior, often rolling or sliding repeatedly down steep inclines in a tobogganing action.

RIVER OTTERS

CLASS	**Mammalia**
ORDER	**Carnivora**
FAMILY	**Mustelidae**

GENUS AND SPECIES   **3 Old World species, including European otter, *Lutra lutra*, and smooth-coated otter, *Lutrogale perspicillata*; 4 New World species, including Canadian otter, *Lutra canadensis* (details below), and giant otter, *Pteronura brasiliensis*; 3 clawless species, including *Aonyx capensis***

WEIGHT
6⅗–33 lb. (3–15 kg)

LENGTH
Head and body: 18–32 in. (46–82 cm); tail: 12–22½ in. (30–57 cm)

DISTINCTIVE FEATURES
Tapered, muscular body; broad head with long whiskers; dark brown, gray or black upperparts; often creamy white throat, chin and chest; tail often drags along ground; well-developed claws; webbed feet

DIET
Fish, amphibians, reptiles, birds and mammals found close to or in water

BREEDING
Age at first breeding: 2–3 years; breeding season: March–April; number of young: 1 to 5; gestation period: 60–70 days; breeding interval: 1 or 2 years

LIFE SPAN
Up to 20 years

HABITAT
Coasts, estuaries, rivers and streams

DISTRIBUTION
Much of Canada and U.S.

STATUS
Locally common

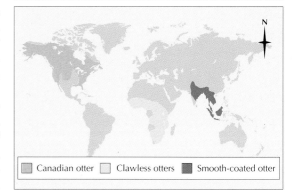

Canadian otter   Clawless otters   Smooth-coated otter

For example, the coat can readily pass from looking sleek and smooth to looking spiny and almost porcupinelike when damp.

Otters do not hibernate. They fish under ice with periodic visits to a breathing hole. It has been said that otters use a trick known among aquatic insects; that is, to come up under ice and breathe out, allowing the "bubble" to take in oxygen from the air trapped in the ice and lose carbon dioxide to the ice and water, and then inhale the revitalized "bubble." This has not yet been proved, however.

Master swimmers

At the surface an otter swims characteristically showing three humps, each separated by 5–8 inches (13–20 cm) of water. The humps are the head, the humped back and the end of the tail curved above the water. When drifting with the current only the head may be in view. Occasionally an otter may swim with the forelegs held against the flanks, the hind legs moving so rapidly as to be a blur. When this is done at the surface, there is a small area of foam around the hindquarters, with a wake rising in a series of humplike waves. The otter also uses this method when submerged, although more commonly it swims with all four legs drawn into the body, which, with the tail, is wriggled sinuously, as in an eel. Leaping from the water and plunging in again, in the manner of a dolphin, is another way in which an otter can gain speed in pursuit of a large fish. Underwater it often progresses in a similar, but smoother, undulating manner.

An otter shows its skill better in its ability to maneuver. It will roll at the surface or, when submerged, pivot on its long axis, using flicks of the tail to give momentum. It can turn at speed in half its own length, using tail and hindquarters as a rudder, or it may swim around and around in tight circles, creating a vortex that brings up mud from the bottom. This tactic may drag up small fish that take refuge under overhanging banks, but this has not been proved.

When an otter surfaces, it stretches its neck and turns its flattened, almost reptilian head from side to side, reconnoitering before swimming at the surface or coming out on land. An otter's eyes are suited to seeing both under and above the water, like a crocodile's.

Otters are territorial and may cover up to 16 miles (26 km) overland in a night; certainly European and Canadian otters are at times far from water. Overland they move by humping the back. They may take a couple of bounds and then slide on the belly for 4–5 feet (1.2–1.5 m). On a steep slope they may glide 40–50 feet (12–15 m). On a muddy or snow-covered slope otters often retrace their steps to slide repeatedly down the slope in a form of play.

Otters live in rivers, especially small rivers running to the sea or to large lakes, and in lakes, estuaries and coasts. They like weed-free rivers that are undisturbed by humans. In parts of Scotland and Ireland river otters live on the coast and should not be confused with sea otters.

The European otter has a varied diet of fish, small invertebrates, particularly crayfish and freshwater mussels, birds, small mammals such as rodents and rabbits, frogs and some vegetable matter. The main prey fish appear to be eels and slow-moving fish, but salmon and trout are preferred if they can be captured.

Cubs soon learn to swim

Mating takes place in water, at any time of the year, with a peak in spring and early summer. After a gestation of about 60–70 days, between one and five cubs are born, blind and toothless, with a silky coat of dark hair. There is uncertainty about when the cubs' eyes open, the only reliable record being 35 days after birth. The cubs stay in the nest for the first 8 weeks of their life and do not leave their mother until just before she mates again.

Young otters swim naturally, as is shown by cubs hand-reared in isolation. The indications are, however, that the mother must coax them or push them into the water for their first swim. In the early days of taking to water, a cub will sometimes climb onto the mother's back, but normally the cubs swim behind their mother. On rare occasions two or more family parties will swim one behind the other. When this happens, a line of humps is seen, and as the leading otter periodically raises her head to take a look around, the procession resembles the traditional picture of the sea serpent.

A startled otter emits a loud cough, but when mating or fighting it grunts or makes a high-pitched noise. Friendly otter "talk" includes chuckling and chirping.

ROACH

Like other members of the carp family, the roach has no true teeth, but some of its gill arches are modified to form pharyngeal teeth in the throat.

THE ROACH IS A common freshwater fish of Europe and Asia. It is one of the numerous species in the genus *Rutilus*, which is placed in the family Cyprinidae, the minnows or carps. Zoologists recognize four subspecies of the roach: *Rutilus rutilus caspicus, R. r. henckli, R. r. fluviatilis* and *R. r. aralensis*. It is a different species of fish from the American roach, a name that usually refers to either the golden shiner (*Notemigonus crysoleucas*) or the introduced rudd (*Scardinius erythrophthalmus*).

The roach is generally deep-bodied and somewhat compressed laterally, although the shape can vary from slender, like the common dace, *Leuciscus leuciscus*, to deeper bodied, like the rudd, one of the roach's closest relatives. In coloration the roach is dull bluish to greenish brown on the back and silvery on the flanks, although older and larger individuals may have brassy yellow sides and some specimens may be almost black. The pectoral fins are tinted red, and the pelvic and anal fins are orange to blood red. The roach can be up to 18 inches (46 cm) long and weigh up to 4 pounds (1.8 kg), but the usual weight is about 1 pound (450 g).

Roach are sometimes confused with other fish. Small European chub, *Leuciscus cephalus*, are sometimes mistaken for large roach, and the rudd is sufficiently like the roach to make distinguishing between the two species difficult. Some differences are that the rudd's fins, including its tail fin, are more red. Its back is brownish olive to green, and there is sometimes a golden sheen on the sides. Rudd can grow a little larger than roach, up to 20 inches (51 cm), and can weigh up to 4½ pounds (2.1 kg). In addition the rudd's belly is keeled in the region of the anal fins, and the iris of the eye is gold, instead of red as it is in the roach. Identifying a roach is made more difficult by the fact that the roach and the rudd hybridize, as does the roach and the common bream, *Abramis brama*. The roach occurs in Britain and Ireland, across Europe north of the Pyrenees and Alps and into Asia, and has been introduced into Australia. It lives in still or running water, provided the current is not too rapid, and tolerates poor-quality, even polluted, water. The roach can also survive in the brackish waters of the Baltic, Black, Aral and Caspian Seas. The form that inhabits these areas of water migrates into rivers to breed.

Bottom feeders

Roach live in shoals. Although they may come to the surface to take insects, they tend to avoid the light and in winter retire to deeper water. The

ROACH

CLASS	**Osteichthyes**
ORDER	**Cypriniformes**
FAMILY	**Cyprinidae**
GENUS AND SPECIES	***Rutilus rutilus***

WEIGHT
Up to 4 lb. (1.8 kg)

LENGTH
Up to 18 in. (46 cm)

DISTINCTIVE FEATURES
Moderately deep body; fairly short head; high dorsal fin, with origin above pelvic fin; large scales; dull bluish or greeny brown on back; sides silvery, sometimes brassy yellow; orange to bright red pelvic and anal fins; other fins gray brown; eye has red iris

DIET
Insects, crustaceans, mollusks and plants, with adults preferring plants

BREEDING
Age at first breeding: 2–3 years (male), 3–4 years (female); breeding season: April–June; number of eggs: 5,000 to 200,000

LIFE SPAN
Up to 12 years

HABITAT
Lowland rivers and lakes, perhaps more common in slow-moving or still, muddy waters; migratory brackish-water populations in Aral, Baltic, Black and Caspian Seas

DISTRIBUTION
Much of Europe, except north and south; introduced to Australia

STATUS
Common

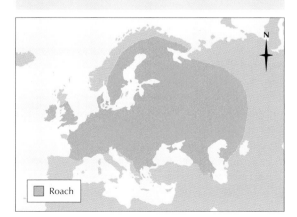

Roach

upper jaw of these fish overhangs the lower, a feature linked with their being mainly bottom feeders. The rudd, by contrast, has a jutting lower jaw, since it is more of a surface feeder. Roach are omnivorous. Besides insects, they take crustaceans and mollusks, although adults seem to prefer plant food.

In the spawning season, male roach develop white nuptial tubercles on the head and body scales. Spawning takes place in April and May, the females depositing the eggs over plants or sometimes gravel, to which they stick, usually in shallow, overgrown stretches of water. The transparent pale to greenish eggs are just over 1 millimeter in diameter and hatch in 9–14 days. The larvae, ⅕ inch (5 mm) long, stay put for 8–10 days, hanging on the vegetation, until the food in the yolk sacs is used up. Then they swim about in dense shoals, sometimes associating with rudd and bream. Growth depends very much on the local conditions. Under the best conditions a roach can reach 3½ inches (9 cm) in 1 year, 5 inches (13 cm) in 2 years and 10 inches (25 cm) in 6 years. Because growth is regular, it is fairly easy to estimate the age of a roach.

Roach predation
Roach larvae are vulnerable to bottom feeders large and small, including the water scorpion, *Nepa cinerea*, and the larvae of dragonflies. The fry are vulnerable to fish and to other predatory insects, such as the great diving beetle, *Dytiscus marginalis*. Mature roach are eaten by European perch, *Perca fluviatilis*, and pike, *Esox lucius*, and even by large rudd. Roach are also a favorite target of anglers, partly because they are numerous but also for the sport they provide: it takes special skill to catch the large specimens.

The roach is not particulary important as a food fish, although it is commercially fished in parts of its range, including Poland.

ROADRUNNER

The greater roadrunner is a nonmigratory bird of the southwestern United States and northern Mexico. It is the state bird of New Mexico.

LARGE, GROUND-DWELLING members of the cuckoo family (Cuculidae), roadrunners are famous for being able to run fast. They have strong legs and zygodactile feet, which means that two toes face forward and the other two backward. The wings are short and rounded, and roadrunners seldom fly.

Roadrunners are found in dry, open habitats, including deserts, of the United States and Central America. There are two species: the lesser roadrunner, *Geococcyx velox*, ranges from Mexico into northern Central America, while the greater roadrunner, *G. californianus*, the subject of this article, is confined to the southwestern United States and northern Mexico. Also known as the chaparral cock, the greater roadrunner is 23 inches (58 cm) long from the bill to the tip of the long tail. It is is brown, streaked with buff and white, with black iridescent feathers in the tail and a short, uneven crest. It has a blue-and-orange streak of bare skin behind the eye.

Rapid runners

Greater roadrunners live in hot, dry country. They survive hot weather by reducing their activity by half during the hottest parts of the day, seeking shade and becoming fully active again only when the air has cooled. Other physiological adaptations to their environment include the abilities to reabsorb water from their feces before they void it and to expel surplus salt via a nasal gland instead of via the urinary tract. Furthermore, their predatory habits enable these birds to obtain much vital moisture from their prey.

Roadrunners rely on running for catching their food and escaping from danger. They were given their name before the days of automobiles, when they used to run alongside horses and carriages. Although a roadrunner was reportedly clocked at a speed of 26 miles per hour (42 km/h) when being chased by a car, reliable measurements put the bird's top speed at 18 miles per hour (29 km/h). It has been calculated that a roadrunner moving at 15 miles per hour (24 km/h) takes 12 steps per second. At speed the roadrunner's feet barely seem to touch the ground, and it uses its wings and long tail to maintain its balance and help it turn. A roadrunner flies only in extreme danger and is unable to remain airborne for long.

Pounded prey

Roadrunners eat a wide variety of small animals, including beetles, spiders, scorpions, grasshoppers, small birds, rodents and lizards. The birds simply snap up small items such as insects from plants, or flush them out first by beating their wings. Roadrunners are also known to follow deer, picking up the insects they disturb. The birds catch larger prey by a quick sprint, seizing the prey and dashing it against the ground to kill it before swallowing it whole. If the prey's body is tough, it is pounded repeatedly until it is reduced to an easily swallowed morsel.

Roadrunners are known for their habit of catching snakes, and they enjoy the same reputation as snake killers as the mongooses (family Viverridae) of Africa and Eurasia. While both roadrunners and mongooses undoubtedly kill venomous snakes, the numbers they kill are often exaggerated. For example, examinations of the stomachs of more than 80 roadrunners showed that 70 percent of the birds' food was insects and only 4 percent lizards and snakes. Nevertheless, the idea of their being ruthless enemies of snakes is deeply rooted and forms the basis of extravagant legends. One is that a roadrunner advances with a prickly cactus leaf held in its bill as a shield and builds up a wall of cactus spines around a sleeping snake. A roadrunner deals with a snake in the same way as does a mongoose. It circles about it, keeping clear

GREATER ROADRUNNER

CLASS	**Aves**
ORDER	**Cuculiformes**
FAMILY	**Cuculidae**
GENUS AND SPECIES	*Geococcyx californianus*

ALTERNATIVE NAME
Chaparral cock

WEIGHT
Up to 13½ oz. (380 g)

LENGTH
Head to tail: 23 in. (58 cm)

DISTINCTIVE FEATURES
Large size; fairly long legs; brown overall with shaggy, streaked appearance; tail long, white-edged and usually held at upward angle; crest short, ragged and often raised; white stripe and patch of blue-and-orange skin behind each eye

DIET
Invertebrates, lizards, snakes, small birds and rodents

BREEDING
Age at first breeding: 2 years; breeding season: eggs laid April–May; number of eggs: 2 to 6; incubation period: 18–20 days; fledging period: about 18 days; breeding interval: usually 1 year, but 2 broods per year in Sonoran Desert, Arizona

LIFE SPAN
Up to 8 years

HABITAT
Dry, open habitats, including rocky deserts, chaparral and grasslands

DISTRIBUTION
Southwestern U.S. and northern Mexico

STATUS
Common

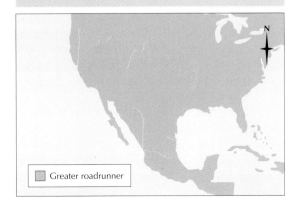

Greater roadrunner

of the snake's fangs using its superior speed and agility. Then, when the opportunity arises, the bird rushes in and strikes at the snake with its pointed bill, stunning the reptile with repeated blows before seizing it, dashing it to the ground and swallowing it gradually, head first.

Wooing with a patter

Little was known about roadrunner breeding until quite recently, and even now scientists' knowledge is based on the behavior of captive birds. Although naturally furtive, roadrunners settle down in captivity and become very tame. At the beginning of the breeding season, the male stakes out a territory, advertising his presence with a song that consists of a series of *coos* descending the scale. Courtship begins with the male offering the female food, which he does not hand over immediately. After presenting this offering he raises his crest, flicks his tail and patters his feet, cackling rapidly at the same time. Then he bows and coos, repeating the whole performance until the female accepts him. Only after mating has taken place does the male part with the food.

The female builds a nest from twigs, which she weaves into a shallow basket in low vegetation. There are usually two to six eggs in a clutch, sometimes more, and they are incubated by both parents for 18–20 days. Incubation begins as soon as the first egg is laid, so the chicks hatch at intervals. At first they are black and almost naked. They are fed by both parents and sometimes swallow lizards as large as their own bodies. They leave the nest when 1 month old.

This greater roadrunner has caught a lizard. Roadrunners belong to the cuckoo family, but unlike some cuckoo species, they build their own nests instead of laying their eggs in the nests of other species.

ROBBER CRAB

As soon as it finds a pool of water, the robber crab is able to immerse itself and thereby quickly restore its blood to its normal concentration.

Robber crabs are able to climb trees, frequently coming down by literally falling to ground level. It is probable that they use this method because it is the easiest way of descending from a tree. They can fall at least 15 feet (4.5 m) without being hurt.

Spawns in the sea

Although completely land-living, the female robber crabs must go to the sea to lay their eggs, from which hatch the typical zoea larvae. These larvae spend some time in the sea and then change to a post-larval form called a glaucothoe. At this stage they migrate onto land and for a while live in empty snail shells, another sign of this crab's close relationship to hermit crabs.

Contradictory reports

Popular belief holds that robber crabs climb trees to wrench off coconuts to eat. The first detailed account of this behavior was given in 1705, and in 1769 the Swedish naturalist Carolus Linnaeus named the crab *Birgus latro* (*latro* meaning robber). In his narrative of the voyage of HMS *Beagle*, the English naturalist Charles Darwin accepted the story of robber crabs opening coconuts but was skeptical about their ability to climb trees. He gave an account of the crab tearing away the coconut husk, fiber by fiber, and always from that end where the eyes of the coconut are situated. Darwin went on to describe how the crab hammers one of the eyes with its claws to make an opening, scoops out the fleshy pulp and eats it. He took this story on trust, as did many other naturalists after him. The story, often even more detailed than Darwin's account, became firmly established and was eventually reported as fact in biological textbooks. Some accounts tell how the crab, having made a hole in one of the eyes, grips it with one of its claws and bangs it against a rock.

In 1939 a Dutch naturalist named Reyne, and later the naturalist Carl A. Gibson-Hill, investigated this issue by keeping robber crabs in captivity with an ample supply of coconuts and nothing else. The crabs did not take to the coconuts, and starved to death. However, other

Robber crabs live on land and are good tree climbers. However, they must return to the sea to breed.

THE ROBBER CRAB, related to hermit crabs, grows up to 1½ feet (45 cm) long and weighs up to around 9 pounds (4 kg). Its walking legs are broad and strong, as are its claws. Behind the usual crab carapace the abdomen is hard and symmetrical, unlike the soft, asymmetrical abdomen of hermit crabs, family Paguridae. The robber crab has abdominal legs on only one side, which suggests it is descended from a hermit crab ancestor that gave up using empty mollusk shells as a shelter and developed a hard abdomen, but could not regain the legs it had lost. Robber crabs range across the Indian Ocean and South Pacific, especially on islands where predators are few.

Tree-climbing land crab

Adult robber crabs live on land and, like other land crabs, have only small gills. However, the walls of the gill chamber are lined with a spongy tissue that is very rich in blood vessels and acts as a lung. Robber crabs drown if they remain submerged for a day or so. This crab is also adapted to life on land in other ways. The shell allows little water to escape, so the crab does not readily dry out. Moreover, the robber crab can withstand a high loss of body water and the concentrated condition of the blood that results.

crabs, given the coconut pulp, readily ate it. This discovery relates to the fact that coconut shells alleged to have been attacked by robber crabs are often seen to have been either gnawed open by rats, to have split on falling from the tree or to have been opened by some other means.

ROBBER CRAB

PHYLUM	**Arthropoda**
CLASS	**Crustacea**
ORDER	**Decapoda**
GENUS AND SPECIES	***Birgus latro***

ALTERNATIVE NAME
Coconut crab

WEIGHT
Up to about 9 lb. (4 kg)

LENGTH
Up to 1½ ft. (45 cm)

DISTINCTIVE FEATURES
Hard abdomen; long, pointed, strong legs; powerful crushing claws

DIET
Mainly coconut flesh and husk, screwpine fruit, turtle hatchlings and dead rats

BREEDING
Separate sexes; female releases eggs in sea; larval period: 4–5 weeks

LIFE SPAN
Up to about 50 years

HABITAT
Open woodland and other coastal habitats on islands; female enters sea to breed

DISTRIBUTION
Islands in Indian Ocean and South Pacific, from Comoros Islands and Seychelles east to Tuamotu Archipelago, French Polynesia

STATUS
Locally common

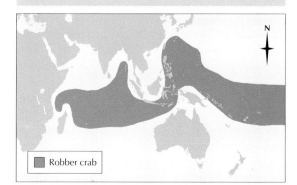

Robber crab

The food of robber crabs, as of many marine crabs, is carrion, along with any coconut pulp it can get without having to break open the shell. Why the crabs should climb trees at all is a matter of scientific conjecture. They may do so in order to drink water trapped in the bases of the leaves. However, Gibson-Hill observed that robber crabs would dehusk coconuts, but not break them, and that they ate fruit and berries and the pith of certain trees, as well as carrion, including the flesh of dead robber crabs. Gibson-Hill claimed that robber crabs are especially fond of the fruit of the tree *Arenga listeri*, and noted that if one crab managed to climb to the top of such a tree, a number of others rapidly collected underneath to catch berries that were dislodged.

In their book *Sable Noir*, published in 1959, Swiss zoologists May and Henri Larsen described how, in the New Hebrides, they watched a robber crab tear the husk from a coconut, carry the nut up a tree and drop it to the ground. The nut fell on soft ground and did not break. Ten times the crab descended, picked up the coconut and dropped it again, until it finally fell on stones and cracked. The crab then came down from the tree and ate the pulp.

Human threat

Robber crabs are highly valued as a food item. They are easily captured by hand after being attracted by a split coconut. With the increasing tourist industry, the value of the crabs has increased. The crab population has declined because of this uncontrolled harvesting and because of changing land-use patterns and other human activities.

Using their massive and powerful claws, robber crabs regularly fight each other to protect their territory.

ROBIN, AMERICAN

The American robin is a common sight in suburban gardens, where it forages for food such as insects and their larvae and earthworms.

THE EARLY COLONISTS IN America gave the name robin to a red-breasted bird that reminded them of the European robin, *Erithacus rubecula* (discussed elsewhere in this encyclopedia), of their homeland. American and European robins are both members of the thrush family, as is suggested by the speckled breast of the young birds, but the American robin is more closely related to the song thrush (*Turdus philomelos*), blackbird (*T. merula*) and fieldfare (*T. pilaris*) than to the European robin. Indeed, the American robin was originally called the fieldfare by some colonists.

The American robin is the largest North American thrush, reaching 10 inches (25 cm). The head, back, wings and tail are brownish gray, the tail being darker and the head dark brown to black, with a black-and-white speckled throat and a white eye-ring. The breast and belly are brick red; the lower body and vent are white.

Distribution and habits

The range of the American robin covers most of the United States, as well as much of Canada. The bird is found far to the north, breeding just beyond the northernmost limit of tree growth, and in the southeastern states its range is gradually extending southward, toward the Gulf of Mexico and the western Atlantic Ocean. While the American robin is more common in deciduous woodland, it is tolerant of climate extremes and may be found anywhere from dense forests to open plains.

The American robin is a migratory bird, the whole population shifting southward during the fall. Consequently, the northernmost robins spend the winter where their more southerly neighbors breed. In the spring, the robins are among the first migrants to return to any area. As a result, they have come to be known as the harbingers of spring.

AMERICAN ROBIN

CLASS **Aves**

ORDER **Passeriformes**

FAMILY **Muscicapidae**

GENUS AND SPECIES **_Turdus migratorius_**

ALTERNATIVE NAMES
Canada robin; northern robin; robin redbreast

WEIGHT
2⅓–3 oz. (65–84 g)

LENGTH
Head to tail: about 10 in. (25 cm); wingspan: 15–16½ in. (38–42 cm)

DISTINCTIVE FEATURES
Largest North American thrush; grayish brown back and wings; blackish brown head and tail; white eye-ring; whitish streaks on throat; brick-red breast and belly; white on lower belly and vent

DIET
Invertebrates; also berries and other fruits

BREEDING
Age at first breeding: 1–2 years; breeding season: April–July; number of eggs: usually 3 or 4; incubation period: 12–14 days; fledging period: 14–16 days; breeding interval: 2 or more broods per year

LIFE SPAN
Up to 11 years

HABITAT
Gardens and woodlands; mountains up to 12,000 ft. (3,600 m) in west of range

DISTRIBUTION
North and Central America, from Canada south to Guatemala

STATUS
Common

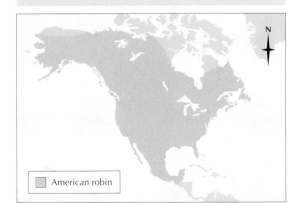

American robin

Although it is primarily a woodland bird, the American robin has adapted its habits to share human environments, and in doing so has become popular with many householders. It often searches for food on garden lawns and sometimes builds its nests in houses and sheds.

American robins feed on a mixture of berries and insects, probably opting for whichever is most readily accessible. An examination of the stomach contents of a sample of American robins revealed that 42 percent of their diet was made up of insects; half of these were accounted for by beetles and half by grasshoppers.

Female builds the nest

Across its range the American robin is one of the first birds to begin laying. Nests are usually built 2–20 feet (0.6–6 m) from the ground, but may be found as far as 80 feet (24 m) up in a tree.

The largest member of the North American thrush family, the American robin is about twice the size of its European namesake.

Newly hatched chicks are defenseless. The parents feed them constantly during their first two weeks, after which time they are able to fly.

Usually there are two broods each year, although rarely there are three. In more northerly areas the nest for the first brood is usually made in coniferous trees, because the deciduous trees that the robin would prefer for the second brood are bare at this time.

The nest is built by the female, with the male assisting only in the collection of material. Even so, he makes fewer trips and carries less material in any trip than his industrious mate. Moreover, if she is busy shaping the nest when he arrives, he is likely to drop his load rather than wait to give it to her.

Building the bowl-shaped nest may take as little as 1 day's activity on the part of the female. There are three stages in its construction. First, the rough outer foundations are laid down, long coarse grass, twigs, paper and feathers being woven into a cup-shaped mass. Then the bowl itself is made out of mud and laid inside the main mass. If there is no readily available source of mud the robin makes her own, either by soaking a billful of dry soil in water, or by wetting her feathers then rubbing them in earth. If there is no hurry and no egg is imminent, the nest construction will stop for a day or two to let the mud dry. Finally, the nest is lined with a layer of soft grasses.

The female may lay from one to six eggs, although three or four is a more usual figure. The eggs are blue green and are incubated for a fortnight, by the female only. She continues brooding the chicks while they are very young. Later she does so only during bad weather and at night. Sometimes her mate helps feed the chicks.

Predators

At one time the American robin was regarded as a game bird in some southern states. Although it is unlikely to have provided much sport, it was often slaughtered in enormous numbers. Today, the robin is protected over its entire range.

Cowbirds frequently parasitize American robins. Domestic cats catch adult birds and young, and the introduced house sparrows often plunder the robins' nests.

Robin in name only

The United States is not the only country where European settlers gave a bird the name robin because of its appearance. In Australia and New Zealand several birds, mainly flycatchers belonging to the genus *Petroica*, are known as robins. There is an Indian robin, *Saxicoloides fulicata*, while the Pekin robin, *Leiothrix lutea*, commonly kept as an aviary bird, is a babbler. The small, dumpy Jamaican tody, *Todus todus*, is called a robin because of its red breast and in various places there are robin-chats, bush-robins, scrub-robins and magpie robins. These represent a diversity of birds, having little more in common than red feathers somewhere on the breast. They were probably all so named because they reminded settlers of their native country.

ROBIN, EUROPEAN

THIS BIRD WAS GIVEN the name robin redbreast as long ago as the 14th or 15th century; later this became shortened to robin. When Europeans settled abroad they frequently used the name for many other, unrelated birds that also had a red breast, as explained under the preceding entry for the American robin, *Turdus migratorius*.

The European robin is a plump bird with comparatively long legs and is about 5½ inches (14 cm) long from the tip of the bill to the end of the tail. It is olive brown on the back, paler on the underparts, and its most conspicuous feature is its bright orange-red forehead, throat and breast. Several subspecies of robins, differing in small details, occur across Europe and into western Asia and southward into northwestern Africa and adjacent islands.

A follower of large animals

In contrast to its pugnacity toward its own kind, the robin has a reputation for extreme friendliness to humans. This has been explained as the natural result of its feeding methods. A robin habitually flies out from cover to take insects, worms and other small invertebrates exposed in the turf by a gardener or kicked up with the leaf litter by the hooves of large animals. This association with large animals is carried further with humans because many people respond by offering the bird food.

The robin lives in thick undergrowth, from which it flies down to pick up food from the ground before retiring to its perch. On the ground it moves in a succession of long hops, carrying its body in a more or less horizontal position. It pauses at intervals in an upright attitude, frequently flicking its wings and tail. Normally the robin flies for only short distances, but when flying farther it has an irregularly undulating flight.

Year-long songsters

Another interesting feature of the European robin, on which several characteristics of its behavior depend, is that both male and female hold territories, and do so for most of the year, either individually or together. Both sing a melodious song throughout the year, except for a waning in June and a short complete break in July. The song is then resumed in late July and early August, at a time when most other songbirds are silent. This coincides with the end of summer, and for many country-dwelling people

The robin exposes its red breast in territorial confrontations with rivals. Juveniles lack this characteristic feature and are usually brown with reddish and tawny marks.

The robin often forages in gardens for food. It frequently picks up earthworms and other invertebrates that are disturbed by the actions of gardeners.

EUROPEAN ROBIN

CLASS	**Aves**
ORDER	**Passeriformes**
FAMILY	**Turdidae**
GENUS AND SPECIES	***Erithacus rubecula***

ALTERNATIVE NAME
Robin redbreast

WEIGHT
½–¾ oz. (14–21 g)

LENGTH
Head to tail: about 5½ in. (14 cm); wingspan: 8–8⅔ in. (20–22 cm)

DISTINCTIVE FEATURES
Small, plump bird; gray-brown upperparts; orange-red breast; off-white belly; light brown tail; fairly long legs

DIET
Insects, earthworms, centipedes, spiders and other invertebrates; also fruits and seeds in winter

BREEDING
Age at first breeding: 1 year; breeding season: eggs laid late March–June; number of eggs: 4 to 6; incubation period: 14 days; fledging period: 13 days; breeding interval: usually 2 broods per year

LIFE SPAN
Up to 6 years

HABITAT
Deciduous or coniferous woodland; woodland edge; agricultural land with trees; suburban parks and gardens

DISTRIBUTION
Widespread in Europe east to Ural Mountains, apart from Arctic fringes; parts of Morocco, Algeria and Tunisia

STATUS
Very common

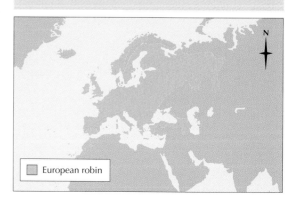

European robin

the robin's song was regarded as an early harbinger of the winter. In the fall the song tends to be softer, with longer phrases.

Since a bird's song is linked with possession of a territory, this extended song period itself suggests that territories are held most of the year. The territorial instinct wanes with the song in June and July; then in August and September males and females stake out their territories once more, and in December and January pairs are formed, usually comprising a male and a female that are neighbors, the territories of the two being joined and shared for the breeding season.

The strong territorial instinct leads to much boundary fighting between robins as well as antagonism when two robins meet outside their territories, for example, at a bird feeder. The first step is a threatening display. One robin points its bill at the sky, fully exposing its red throat and breast, and slowly sways from side to side, as if waving a red flag. If this is ignored by the opponent, a fight, usually no more than a chase, follows. Through experiments, scientists have

established the importance of the red patch. If a tuft of red feathers is wired to a twig, a robin will first display at them and then attack.

Mainly insectivorous

Robins eat mainly insects but also other small invertebrates, such as earthworms, centipedes, millipedes, spiders and wood lice. The birds take small seeds as well as soft fruits at times and berries in the fall. In a fruit garden robins are not a great nuisance because, being territorial, with each territory up to 1 acre (0.4 ha) in extent, they do not attack a crop in numbers in the way that tits and thrushes do.

The threatened sex

The courtship of robins begins with a fight, as is the case in many birds, but it is particularly necessary in the case of robins because the male and female look alike and can be identified only by their behavior. The male robin displays aggressively, the female responding with a submissive posture. Later, the male gives the female a gift of an insect grub as part of the courtship ritual. The hen alone builds the nest, which consists of dead leaves and moss lined with hair and some feathers, but she continues to

be fed by the cock, fluttering her wings at him like a juvenile. This takes place from the end of March onward, the nest typically being built in a hole in a bank or rotten tree stump. However, robins are capable of adapting a wide variety of sites to their nesting needs, including old abandoned kettles and empty cans. Robin nests have even been found under the hood of a car.

There are usually four to six white eggs with many reddish freckles, which are incubated by the hen alone for 14 days. She is fed on the nest by the cock, or is called off by him to take food. Fledging takes another 13 days, the young being fed by both parents. There may be two or, less commonly, three broods a year. The newly fledged juvenile does not have the adults' red marking. It is spotted like a young thrush or common redstart, *Phoenicurus phoenicurus*.

Robins have much the same predators as other small songbirds, but probably suffer less than many because they spend much of their time hidden in undergrowth. Traditionally, they have been spared persecution by humans because of beliefs, widespread in Europe and taking a variety of forms, that bad luck, even disaster, will follow the killing or caging of a robin, or even interference with its nest.

The male robin brings food to the female during the nesting period. The female usually remains on the nest, which she constructs on her own.

ROCK DOVE

Feral descendants of the rock dove are found in towns and cities the world over.

THE ROCK DOVE, OR ROCK PIGEON, is the ancestor of the domestic pigeon and also of the feral pigeon, which has reverted to the wild and is a familiar sight in many towns. There is essentially little difference between pigeons and doves, both of which belong to the family Columbidae. Domestic and feral rock doves exist in a wide variety of forms. Their colors range from almost black to almost white, with a greenish or pinkish tinge to the neck, while some subspecies are checkered. The domestic varieties, in particular, frequently have abnormal skull and bill structures. Pigeons and doves have a very limited range of calls, most of which are variations on a cooing sound.

The descendants of rock doves may be divided into different categories. Rock doves themselves are truly wild birds. By contrast, racing or homing pigeons are domestic pigeons that are kept for their ability to return home when they are released at a distance. They are bred for their speed as well as their homing abilities and may attain average flight speeds of 44 miles per hour (70 km/h). So-called fancy pigeons are domestic pigeons that have been bred exclusively for their appearance, which is frequently spectacular. Feral pigeons are pigeons that live in the wild but have descended from dovecote or domestic pigeons. They are often seen in towns and cities.

Rock doves are 12–14 inches (30–35 cm) long, far smaller than wood pigeons, *Columba palumbus*, and are further distinguished from them by their darker plumage, with no white on the wings.

In most places it is now impossible to distinguish wild rock doves from feral pigeons. Rocky cliffs where wild rock doves became extinct many years ago have since become colonized by feral pigeons in some places. Since the two birds can look identical, it is virtually impossible to establish their ancestry. In all but the most remote places wild stock and feral birds have interbred.

Wild rock doves range from the Faeroes, 200 miles (320 km) north of Scotland, to India and Sri Lanka and south to the Sudan. In many places the birds were either introduced by humans, as in the Azores and the Faeroes, or followed humans of their own accord.

ROCK DOVE

CLASS	**Aves**
ORDER	**Columbiformes**
FAMILY	**Columbidae**
GENUS AND SPECIES	***Columba livia***

ALTERNATIVE NAMES
Rock pigeon; feral pigeon; domestic pigeon; racing pigeon; fancy pigeon

LENGTH
Head to tail: 12–14 in. (30–35 cm)

DISTINCTIVE FEATURES
Medium-sized blue-gray pigeon; 2 black bars across rear half of inner wing; most subspecies have white rump

DIET
Seeds, buds, invertebrates and human garbage

BREEDING
Age at first breeding: 1 year; breeding season: eggs laid year-round depending on region; number of eggs: 2; incubation period: 16–19 days; fledging period: 35–37 days; breeding interval: 5 or more broods per year

LIFE SPAN
Not known

HABITAT
Rocky cliffs; open country; feral populations in human settlements, including large cities

DISTRIBUTION
Large range, including all continents except Antarctica

STATUS
Common to abundant

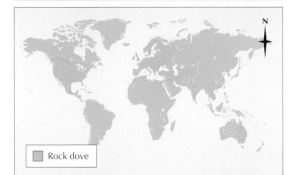

Rock dove

Declining doves

Rock doves live in pairs or small parties but form flocks where they are particularly numerous. They do not migrate, but may move to colonize new areas. In some places rock doves are declining, partly through persecution by humans and partly through competition from feral pigeons. The wild stock is also diluted through interbreeding with feral pigeons.

Rock pigeons are persecuted for several reasons. Both adults and nestlings have been taken as food for centuries, but now they are sometimes trapped for fear that they will interbreed with fancy pigeons, and like feral pigeons they are also shot by farmers in the effort to protect crops.

The pigeons that are so common in towns and cities are feral pigeons, or feral rock doves. They are now almost completely dependent on humans for food and shelter. People frequently give feral pigeons food in public places, but to the authorities the birds are a nuisance. Pigeons foul buildings with their droppings, making expensive cleaning operations necessary, and when they roost or nest inside warehouses the droppings may contaminate food. Pigeons may constitute a menace to public health, since they

Wild rock doves live in open country, usually nesting on rocky outcrops or cliffs. Their original home may have been in semideserts, as they rarely settle on trees.

carry diseases that can infect humans. There is, however, no proof as yet that any diseases have been transmitted from pigeons to humans.

Apart from humans, pigeons that live in towns face fewer threats than those that live in the countryside. This is probably the reason for the wide variety of plumage in towns, especially of black plumage, for town-dwelling pigeons are not subject to the same pressures of natural selection that the country pigeons face.

Seed eaters

Rock doves feed mostly on seed, which they pick off the ground. In cultivated areas they feed on crops, such as grain, potatoes and peas. Seeds of wild grasses and other plants are also taken, along with some animal food such as earthworms and their egg cocoons, snails and slugs. Their diet is sometimes inadequate and town pigeons have been known to peck at crumbling mortar in order to obtain the necessary lime for producing eggshells.

Rapid turnover

In the wild, rock doves nest in caves or in rock crevices, but in towns they regularly make use of buildings, coming in through holes or windows to nest under roofs. Nesting takes place all year round, but few doves lay in the winter months.

The rock dove's courtship is similar to that of the wood pigeon. The male runs around the female and bows in front of her with his neck puffed out and his tail spread. At the same time, he coos continuously. The two eggs, sometimes only one, are laid on a nest of grass stems, roots, twigs or even seaweed. The male collects the material for nest construction, but it is the female that makes the nest.

Incubation lasts 16–19 days and is shared by both sexes. The chicks, or squabs, are tended for 5 weeks. At first they are fed on pigeon's milk by both the male and female. This is produced in fluid-filled cells that line the pigeon's crop, a thin-walled, saclike food-storage chamber. The crop tends to be particularly well developed in pigeons and doves. Pigeon milk is highly nutritious and contains more protein and fat than cow or human milk. It is the exclusive food of the squabs for several days after hatching, ensuring that they receive an adequate supply of nutrients to sustain the high growth rates typical of pigeons. Later the milk is supplemented with solid food. A second clutch may be started before the first brood of young is independent, and for about two weeks the parents may have both eggs and young to care for.

Peregrine predators

Young rock doves are safe from most predators while they live in caves or buildings but jackdaws sometimes rob the nests. The main predators of adults are peregrine falcons, which have been persecuted for attacking homing pigeons and other domestic variants.

Rock doves have a global distribution and are found on every continent. The rock dove below was photographed in India.

ROCK WALLABY

Rock wallabies are sometimes called the Australian chamois because of their rocky habitat and leaping abilities. There are several species, some of them not much larger than hares. They are 2–3 feet (60–90 cm) high with cylindrical tails, 16–28 inches (40–70 cm) long, that are tufted at the tip in some species and not thickened at the base as in kangaroos.

The rock wallabies' long, thick coat is gray, sandy or brown, the underparts being pale cream, beige or yellow. Some species have stripes, patches or other markings. The ring-tailed rock wallaby, *Petrogale xanthopus*, has a series of dark rings along the tail. Also known as the yellow-footed rock wallaby, it is one of the most colorful, being gray with yellowish markings on the limbs and tail. Rock wallabies are found locally in areas of rocky outcrops and boulders in all parts of Australia.

Nimble marsupials

Naturalist John Gould studied Australian fauna in the early 19th century and compared rock wallabies to monkeys. They are extremely agile among rocks and can scale leaning trees at top speed, making huge leaps. They grip the branches with the soles of their feet and the two central toes, which bend more freely than the toes of other wallabies. The soles of the feet are well padded and edged with hair, and the granular skin gives a nonskid surface.

Rock wallabies scamper over rocks with ease, and in places the rocks are polished to the smoothness of glass by the feet of generations of rock wallabies. They can leap across chasms up to 13 feet (3.9 m) wide, dash for cover into a crevice judging the width perfectly and climb almost vertical rock faces, using only the hind feet with the tail as a balancing organ.

They avoid the hottest part of the day by sheltering among the rocks, and come out in the early morning and late afternoon to feed and sunbathe. They may go fair distances from their rocky homes to feed and can be seen traveling over the flat ground moving awkwardly, their body being held more or less horizontal with the tail curving upward. The tail is never used to support the body as in other wallabies. When alarmed, rock wallabies thump on the ground with the hind feet, alerting their fellows.

Living without drinking

Although usually said to be nocturnal, Vincent Serventy reports that where they are not persecuted rock wallabies can be seen out and about at almost any time of the day, except when the sun's heat is fierce. Apparently even this varies with the species, and the short-eared rock wallaby, *Petrogale brachyotis*, in the northwest of the state of Western Australia has been reported as standing up to temperatures of 136° F (58° C) among the sandstone rocks. Rock wallabies eat mainly grass, which is often lacking during the dry season. Then they turn to leaves, bark or roots. It seems that rock wallabies can go for long periods without drinking when the pools among the rocks dry out, getting their moisture from the more juicy barks and the roots.

Vulnerable to imported enemies

Their breeding habits are like those of other wallabies, the females having a single pouched young. The joeys are especially vulnerable to eagles and to carpet snakes—the large Australian python. The adults have few natural enemies, apart from dingoes, which is probably why they do so little to defend themselves. Their reaction to danger is to dash into the security of rocks and lie low, raising their heads occasionally to see what is going on. Such tactics make them very vulnerable to domestic dogs and introduced cats and foxes. They have suffered most, however, by being shot for their pelts, the fur of the ring-

The brush-tailed rock wallaby, Petrogale penicillata, is one of 14 species of rock wallabies with their characteristically long, cylindrical tails.

The yellow-footed or ring-tailed rock wallaby is surefooted on rocky outcrops because its hind feet have padded soles.

tailed rock wallaby being especially valued. The rock wallabies have received legal protection in some places, but all species have had their numbers reduced in some parts of their range, and some are in possible danger of extinction if sanctuary is not provided.

It is not clear from the available records whether foxes were taken to Australia for hunting or to help keep down the rabbit population. Whichever it was, it is probable that there were plenty of rabbits already running wild when the foxes began to spread. The usual remark made by writers on the subject is that the foxes preferred the flavor of marsupial flesh to that of rabbits. It is more likely that the foxes took more of the native marsupial herbivores not because they taste better but because they are less wary and therefore easier to catch than rabbits. Whatever the reason, the result has been that several of the native species have been dealt a severe blow from this cause alone, rock wallabies being badly affected.

Conveyor-belt teeth

All the rock wallabies belong to one genus, *Petrogale*, except the little rock wallaby, *Peradorcas concinna*. It is about half the size of the others and has similar habits. Its teeth, however, are very different. It has nine molars, an unusually high number for a wallaby, but there are never more than five in use at one time. The teeth erupt through the gums successively, and those in front fall out as they are worn down. They are replaced by the molars behind gradually moving forward, rather like a conveyor belt. Such an arrangement is found in only two other mammals, the elephant and the manatee.

ROCK WALLABIES

CLASS	**Mammalia**
ORDER	**Diprotodonta**
FAMILY	**Macropodidae**

GENUS AND SPECIES **15 species, including ring-tailed rock wallaby, *Petrogale xanthopus* (details below); brush-tailed rock wallaby, *P. penicillata*; and little rock wallaby, *Peradorcas concinna***

WEIGHT
6½–20 lb. (3–9 kg)

LENGTH
Head and body: 19½–31½ in. (50–80 cm); shoulder height: 2–3 ft. (60–90 cm); tail: 16–28 in. (40–70 cm)

DISTINCTIVE FEATURES
Thick dense fur; gray, sandy or brown upperparts; pale cream, beige or yellow underparts; black rings on tail; padded soles on feet; long, cylindrical tail

DIET
Mainly grasses, bark and tree sap; can go for long periods without drinking

BREEDING
Age at first breeding: 18–20 months; breeding season: year round; number of young: usually 1; gestation period: 30–32 days; breeding interval: 1 year

LIFE SPAN
Up to 10 years or more in captivity

HABITAT
Rocky scrub and boulders in hills and mountain ranges

DISTRIBUTION
Throughout much of Australia

STATUS
Near threatened

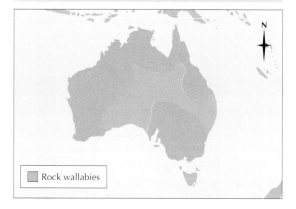

Rock wallabies

ROCKY MOUNTAIN GOAT

THE ROCKY MOUNTAIN GOAT is not a true goat but one of the so-called goat-antelopes closely related to the chamois, *Rupicapra rupicapra*, of Eurasia (discussed elsewhere). With its short neck, large head and pure white coat, it looks more like a domestic animal than a wild species. Its shoulders are 31½–47¼ inches (0.8–1.2 m) from the ground, and the back slopes down to its shorter hind legs. The head-and-body length is 4¼–5⅓ feet (1.3–1.6 m), with the tail just 4–8 inches (10–20 cm) long. It weighs 110–310 pounds (50–140 kg), males being much heavier than females. Both sexes have horns and beards, the black horns being upright but curving slightly backward. The horns rarely grow more than 8 inches (20 cm) in length. The coat of this mountain goat is white or yellowish white, with the hair being long and stiff along the midline of the neck and shoulders, forming a hump. The underfur is thick and woolly.

This North American mountain goat lives in the Rockies and the eastern Yukon border, south to western Montana, central Idaho and northern Oregon. It was introduced into the Black Hills of South Dakota in 1924.

Surefooted bovids

Rocky Mountain goats favor steep mountain-sides and cliffs. Their split hooves help them to move easily along the rock faces, each hoof having a hard, sharp rim enclosing a soft inner pad. They work equally well over slippery ice.

The goats move quite slowly unless alarmed, grazing on grasses, sedges and lichens, and browsing on any available shrubs and tree foliage. In summer they are known to travel several miles to reach salt licks. Although they are agile, mountain goats are unable to work their way through deep snow, so they descend to lower areas in fall where the snow is not too deep. They grow thick winter coats and find food on lower mountain pastures, usually south- or west-facing slopes, or in protected thickets. The seasonal migration may involve a trek of 7 miles (11 km) or more. Along the coastal part of their range deep snow forces certain populations down to sea level. Sometimes the mountain goats avoid floundering in snow by retreating into caves for shelter. In the following spring, with the melting of the snows and new growth of plants and shrubs, they head upward once more.

The mountain goat has several predators, including pumas, wolves, and brown and black bears, as well as golden eagles, which sometimes kill kids. However, they are more likely to fall victim to snowdrifts, which deprive them of food, or to avalanches and rock slides, which sweep them to their death. They are also hunted in some areas by humans for food or game.

Active kids from birth

The mating season for these goat-antelopes occurs from November to January. Although the males make a great show of fighting, they do not get into serious combat over the attentions of a particular nanny (female) goat. The young are born about 180 days after mating. Usually they number one or two, although very occasionally a litter of three may be born. The young appear to be able to stand within 10 minutes of their birth and nurse within 20 minutes. They weigh about 7 pounds (3 kg) and stand about 31 inches (78 cm) high. Kids willingly follow their mother across precipitous mountain slopes within a week of their birth. The females and young stay together in small herds throughout the year, the males joining them only for the summer and mating season. The males live alone or in small bands of two or three when they are not part of a summer breeding group.

Adult Rocky Mountain goats have thick, shaggy coats to keep them warm. An undercoat of dense, finer fur forms an effective insulating layer against the cold.

Within a few days of birth, kids can follow their mothers over the most difficult terrain. The young grow so quickly that at just 3 months they weigh 40 pounds (18 kg).

ROCKY MOUNTAIN GOAT

CLASS	**Mammalia**
ORDER	**Artiodactyla**
FAMILY	**Bovidae**
GENUS AND SPECIES	***Oreamnos americanus***

WEIGHT
110–310 lb. (50–140 kg)

LENGTH
Head and body: 4¼–5⅓ ft. (1.3–1.6 m); shoulder height: 31½–47¼ in. (0.8–1.2 m)

DISTINCTIVE FEATURES
Thickset body; muscular legs; black eyes and nose; black, curved horns; beard (both sexes); summer coat: short, woolly, white; winter coat: shaggy, dense, yellowish

DIET
Grasses, lichens, mosses, herbs, woody shrubs and trees

BREEDING
Age at first breeding: 30 months; breeding season: November–January; number of young: usually 1; gestation period: 175–185 days; breeding interval: 1 year

LIFE SPAN
Up to 18 years

HABITAT
Steep mountainsides and cliffs

DISTRIBUTION
Rocky Mountains, south to northern Oregon

STATUS
Locally common

Rocky Mountain goat

Life among the peaks

The appearance of a mountain landscape, with its bare ridges, high rocky cliffs and deep ravines and valleys, suggests an inhospitable environment. However, many species of animals, often in large numbers, can be found living in these high places, even though a mountain habitat is one of the harshest environments on Earth.

The conditions that an animal faces during the course of a year are extreme. There are high winds, severe cold, extremes of temperature, deep snow and ice and floods at times, with dry periods at other times. The air becomes more rarefied the higher the animals live, and accessing sufficient oxygen can be a demanding task. Wild sheep and ibex range as high as 19,000 feet (5,790 m) in Tibet, wolves and foxes can be found in many mountain ranges at a slightly lower altitude, and yaks manage to live at even higher levels. At such heights the oxygen pressure is less than half that at sea level. As well as coping with the demanding atmosphere, mountain animals must also be capable of finding enough food and water to survive.

All hooved mountain mammals have an excellent coordinated system of muscles, tendons and bones that allows them to leap along mountain slopes without any apparent difficulty.

ROE DEER

THE ROE DEER IS NOT often seen, despite its widespread distribution in Europe and Asia and its large population numbers, particularly in some countries. This is because it is crepuscular, meaning that it is active mainly around twilight, and also because it can move stealthily through dense cover.

A well-grown buck stands up to 3¼ feet (1 m) at the shoulder, weighs about 110 pounds (50 kg) and may be up to 5 feet (1.5 m) in length. The doe is smaller, weighing up to 46 pounds (21 kg). In summer the short, smooth coat is bright reddish brown with white underparts. The deer has a white chin and a white spot on each side of the dark muzzle. In winter the coat becomes long and brittle and the color changes to a dark grayish fawn with a very conspicuous white tail patch. The roe deer has a very short tail, and its ears, with long hairs and whitish insides, are relatively large compared with those of other species of deer. The antlers, which are knobbed at the base, are only ¾–1¼ feet (23–38 cm) long, nearly upright and seldom have more than three tines (prongs). They are shed in November and December, and the new antlers are fully grown and clear of velvet by the end of April or the beginning of May.

Two species today

Once regarded as one species, roe deer are now classified as two: *Capreolus capreolus* and *C. pygargus*. The first of these species lives throughout Western Europe, from Mediterranean lands in the south to southern Scandinavia in the north, and ranges into Iran and northern Iraq. It has been introduced into the United States. *C. pygargus* is found in Siberia, China and Mongolia.

Although it is absent from Ireland, in Britain the roe deer appears to have been formerly the most widely distributed native deer. It seems to have been driven farther north by the increase in human settlements in the south. As a truly wild animal, the roe deer had almost disappeared from England by the latter part of the 18th century, clinging on in Britain only in the Scottish Highlands. However, it was soon reintroduced into woodlands in widely separated parts of the country, and by the mid-19th century had expanded its range south to the Scottish

Borders. Today there are probably more roe deer in the British Isles than ever before, especially in the woods and parks of southern England.

Skulking under cover

Roe deer are skilled at being able to move about almost noiselessly. They are small enough to take advantage of any cover and to skulk in the undergrowth unseen, even in the daytime. They can live in almost any area that has enough cover, but they prefer open woodland, sparsely wooded valleys and low mountain slopes. They are also found in parkland, agricultural land and scrub. Roe deer are good swimmers and often cross rivers, probably to reach new feeding grounds, but they have been known to swim across wide stretches of water, such as the Scottish lochs, and even arms of the sea, for no discernable reason. In moderately populous areas of southern England, roe deer are some-

Roe deer fawns retain their spots only for their first year of life.

The reddish brown coat, grayish face and black band running from nose to mouth is typical of the roe deer's summer coat. Its relatively simple antlers indicates this deer is a young adult.

ROE DEER

CLASS	**Mammalia**
ORDER	**Artiodactyla**
FAMILY	**Cervidae**
GENUS AND SPECIES	***Capreolus capreolus;*** ***C. pygargus***

WEIGHT
33–110 lb. (15–50 kg)

LENGTH
Head and body: 3–5 ft. (0.9–1.5 m); shoulder height: 2–3¼ ft. (0.6–1 m); tail: ¾–1½ in. (2–4 cm)

DISTINCTIVE FEATURES
Short coat of reddish brown fur; darker coloring around ears; white underparts; short, white tail; throat and rump grayer in winter; antlers ¾–1¼ ft. (23–38 cm) in length

DIET
Grasses, herbs, fresh leaves and shoots; also berries, fungi, crops and ornamental plants

BREEDING
Age at first breeding: 16 months; breeding season: July–September; number of young: usually 1 or 2; gestation period: about 165 days; breeding interval: 1 year

LIFE SPAN
Up to 12 years

HABITAT
Woodland, parkland, agricultural habitats, suburbs and scrub

DISTRIBUTION
***C. capreolus*: western Europe east to Iran and northern Iraq; introduced to U.S.**
***C. pygargus*: Siberia, Mongolia and China.**

STATUS
Common

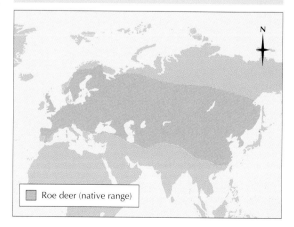

Roe deer (native range)

times seen in gardens in the very early morning, but it is more usual to find only their numerous hoofprints, or slots, in soft ground.

Both the buck and doe have an alarm bark very similar to that of a dog. In the breeding season the doe utters an incessant squeaking that is hardly audible to humans at more than a few yards' distance. A third type of cry, a quavering, high-pitched scream, indicates fear or anguish.

Young shoots preferred

Roe deer feed mainly at dusk and dawn, when they graze on open grassland. Apart from grass, their food ranges from the foliage of broad-leaved trees and shrubs to yew and pine shoots, heather and juniper, as well as briar, bramble and privet. Berries, fungi and clover are also eaten. The deers' habit of eating the tips of young tree shoots often causes widespread destruction of new plantations. They also feed on crops and ornamental plants and are so numerous in some countries that they can be a pest.

Roe rings

One notable feature of the roe deer's rutting season is the use of roe rings, within which a form of courtship takes place. A piece of open ground around a tree, bush or clump of vegetation is chosen, and the buck marks this area out by fraying the young trees around, stripping the bark from them and scraping the ground, sometimes marking the scrapes with scent from glands on his forehead. The tree or bush acts as a central point around which the buck chases the doe, their hooves wearing a ring or figure eight in the ground. Although the buck, which follows the doe closely, seems to be driving her, she is almost certainly a willing partner. The rutting season is July–September and mating takes place within the ring or figure eight.

Implantation of the embryo is delayed and takes place in late December. In the following May or June the doe retires deep into a covert, where her one or two, rarely three, young are born. Their coats are spotted only for the first year. When the fawns are about two weeks old the mother brings them out into more open areas so that they may join the buck. The first antlers that the young have are simple, unbranched prongs that begin to grow in February of the following year; they are fully developed and cleaned by June. In the third year the antlers are forked, with a short tine pointing forward. The antlers of the fourth year have an additional tine that is directed backward. This marks the full extent of antler growth.

Roe deer live in small family groups consisting of a buck, a doe and young. These units break up at the end of the winter, when the young are usually driven away, although sometimes they stay with their mother for several years.

Foxes a danger

Apart from humans, adult roe deer have few predators except in the few areas where wolves and lynx have not yet been eliminated. The fawns, however, are often killed by foxes. More young fall victim to foxes than is generally supposed, for although the does defend their young strenuously, they have great difficulty defending two at once. This is one reason why often only one of a pair of twins survives.

A threat to young trees

The roe deer population stands at several million. In some countries the animal has become a pest, particularly in areas of maturing woodland. In about February the roebuck joins an already pregnant doe and establishes a territory of considerable size, which he vigorously defends against any intruder. Some of these combats may end in death for one, or sometimes both, of the combating bucks.

When two roebucks face each other across a mutual boundary, it is likely that instead of fighting or fleeing they indulge in a displacement activity as an outlet for their aggressiveness and superfluous energy, attacking the nearby trees and scraping their antlers back and forth on the trunks. This activity can be more disastrous to growth than the deer's habit of eating young shoots. It has been found that selective killing of the weaker bucks not only improves the roe deer stock but also guarantees that an area will hold the minimum roe deer population, ensuring that the trees suffer less damage.

Roe deer range widely across Eurasia, from western Europe to the eastern coast of China. The deer below were photographed in Sweden.

ROLLER

ROLLERS ARE COLORFUL BIRDS and are near relatives of bee-eaters and kingfishers. There are 11 roller species, 8 in the genus *Coracias* and 3 in the genus *Eurystomus*. *Coracias* species take their prey from the ground, after a flight from their perch, while *Eurystomus* species capture their prey in the air. Rollers are strongly built and jaylike, ranging from 9 to 13 inches (22.5–32.5 cm) in length with long, shallowly curved bills that have slightly hooked tips. The tail is long, either square or forked at the tip, and the wings taper to a point. Closely related to rollers are the cuckoo rollers and ground rollers of Madagascar, off Africa in the Indian Ocean.

Rollers are confined to the Old World, most of them living in Africa. Their plumage is largely made up of bright blues, greens and reds, and the Abyssinian or Senegal roller, *C. abyssinicus*, is often regarded as the most striking bird in West Africa. Although its back is brown, the rest of its plumage is made up of blues: dark blue on the wings and greenish blue on the body. The two outer tail feathers extend 6 inches (15 cm) beyond the rest of the tail. The Eurasian roller, *C. garrulus*, breeds in North Africa, western Asia and Europe, as far north as Sweden. It has a mainly greenish blue plumage with a chestnut back. The broad-billed rollers are rather smaller than the others. The broad-billed roller of trop-ical Africa, *E. glaucurus*, has a short, triangular yellow bill that has a very broad base. Its back is brown with bright blue on the flight feathers and tail and violet underparts. The dollarbird, *E. orientalis*, so called because of the white patches on its wings, ranges from northern India to the Solomon Islands and Australia. The broad-billed roller and the dollarbird are far more compact and much less slender than other rollers, many of which have elongated tail feathers.

Changing distribution

Rollers are usually solitary and can be seen perching on exposed places such as bare boughs, buildings and telegraph wires or hopping on the ground. Several species are migratory and they travel and spend the winter in small flocks. The Eurasian roller migrates south to tropical and southern Africa for the winter, and the dollarbirds of Australia migrate northward to New Guinea and Indonesia. During the return journey some wander over to New Zealand.

The distribution of the Eurasian roller has been changing over the last century. Once quite common in Sweden, today only a handful of pairs breeds in one locality. At the same time, its visits to the British Isles have become rarer. It is thought that this decrease is due to the wetter summers experienced by northwestern Europe in recent years, which have resulted in reduced numbers of the large insects on which rollers feed. By contrast, in eastern Europe, the population has increased and spread as the result of extensive reforestation, providing the birds with a more suitable habitat.

Opportunistic feeding

Many rollers feed on large insects such as grasshoppers, beetles, butterflies and moths. Having caught prey, the birds then take it to a perch, against which they beat it. Some rollers catch small lizards, birds and frogs or even snakes and scorpions. Others hunt around forest or bush fires, catching the insects and other small animals flushed out by the flames and smoke. Dollarbirds sometimes hunt in the evening in the company of bats.

First feathers are wrapped

Rollers nest in holes in trees and banks or in crevices between rocks and in walls. They often take over and enlarge

One of the most commonly seen rollers, the lilac-breasted roller perches on prominent branches, bushes, wires and fence posts to spot its prey.

LILAC-BREASTED ROLLER

CLASS	**Aves**
ORDER	**Coraciiformes**
FAMILY	**Coraciidae**
GENUS AND SPECIES	***Coracias caudata***

LENGTH
Head to tail: 14–15 in. (36–38 cm)

DISTINCTIVE FEATURES
Slim, rather elongated body; heavy black bill with strongly hooked tip and rictal bristles (stiff feathers) at base. Adult: brilliant greenish blue crown, belly, tail and wings; deep blue flight feathers; white chin and supercilium (stripe over eye); bright lilac throat and breast; tawny brown back; 1 pair of very long outer tail feathers. Juvenile: much duller, with no elongated tail feathers.

DIET
Large insects, such as locusts and grasshoppers; also lizards, scorpions and large spiders

BREEDING
Age at first breeding: probably 2 years; breeding season: September–December (southern Africa); number of eggs: 2 to 4; incubation period: 18–20 days; fledging period: probably 28–32 days; breeding interval: 1 year

LIFE SPAN
Probably up to 10 years

HABITAT
Dry savanna woodland and thornbush

DISTRIBUTION
Eritrea and Ethiopia south through East Africa to northeastern Namibia and northern South Africa

STATUS
Common (southern Africa); fairly common (Kenya, Uganda and Tanzania)

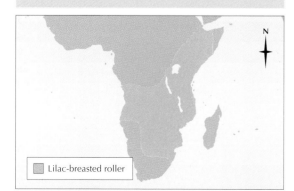

Lilac-breasted roller

Rollers (Indian roller, Coracias benghalensis, above) typically have colorful plumage and a strong, hook-tipped bill. They are named for their aerobatic display flights.

abandoned woodpecker holes, and also use the nests of magpies and pigeons. The white glossy eggs, two to five in number, are laid on the floor of the nest. If there is any lining, it is very scanty. The sexes both incubate the eggs, but the female takes the larger share of the duty.

The chicks hatch out in 18–20 days. At first they are brooded by the female, the male bringing food, which he passes to the hen through the nest entrance. When the chicks are ready to leave the nest in 28–32 days, their developing feathers are encased in protective waxy sheaths. As with related birds, the sheaths burst just before the fledglings leave the nest.

Aerial acrobats

Rollers derive their name from the spectacular tumbling and rolling flight that they perform during courtship. During this display their vivid colors are shown off to the best effect, and at the same time the rollers scream loudly. They also soar skyward and then plummet earthward on closed wings. The precise function of these displays is not known, but sometimes as many as half a dozen rollers display together. Broadbilled rollers do not indulge in this tumbling display, but make long swooping flights.

The rollers' acrobatic skill is also shown when they chase hawks and other large birds that come near their nests. In fact, rollers will even attack humans and dogs that approach too close to their eggs or young.

ROOK

A recently fledged juvenile rook begs for food from its parent, which has a throat pouch full of food (visible beneath the adult's bill).

THE ADULT ROOK CAN BE distinguished from other members of the crow family by the whitish skin at the base of its pointed bill. However, in its first year the rook has black feathers overlapping the base of its bill, as in the carrion crow, *Corvus corone*. The rook is also about the same size as the carrion crow, being 17–19 inches (43–48 cm) long, but its plumage has a blue or purple sheen, whereas the carrion crow's is more greenish. The upper leg of a rook is feathered so the bird looks as if it is wearing short, baggy pants.

The rook's range includes most of Europe except the extreme north. It is also found in much of Asia, where there are two subspecies that are slightly smaller than the European rook.

Strutting rooks

Rooks are lowland birds and are especially numerous where the land is tilled. They feed in large flocks that spread out over the fields and at night gather in communal roosts in trees. On the ground they have a somewhat strutting walk but they also hop, using half-opened wings. The best-known of their calls is the caw, which they utter with the body and head held horizontally and the tail erect and fanned. Rooks have other more musical notes but these are soft and are not audible at a distance. They can be heard, for example, when a particularly bold wild rook accepts food thrown down for it.

Grain and insects

There have long been differences of opinion about rooks' food. Some people claim the birds are beneficial to agriculture by eating so many insects, especially wireworms, whereas others maintain they eat too much grain. In one survey, David Holyoak found that their main food is fallen grain, earthworms and insects. Potatoes, root crops, peas, the seeds of wild plants, acorns, beech mast, walnuts, birds' eggs, snails, spiders and centipedes are also eaten.

Rooks spend much time each day digging with their bills, stabbing the earth obliquely with the bills slightly open. They also hoard surplus food using the same digging action to make a small hole. After placing the food in it, they cover it with earth or grasses. Rooks will bury pine cones or small colored or bright objects in the same way, but the reasons for this behavior are not yet understood.

ROOK

CLASS	**Aves**
ORDER	**Passeriformes**
FAMILY	**Corvidae**
GENUS AND SPECIES	***Corvus frugilegus***

WEIGHT
12–18¾ oz. (340–532 g)

LENGTH
**Head to tail: 17–19 in. (43–48 cm);
wingspan: about 3 ft. (90 cm)**

DISTINCTIVE FEATURES
**Adult: bare whitish face, making pointed bill
appear longer; all-black plumage with blue
or purple gloss; feathered "pants" on upper
legs. Juvenile: black face.**

DIET
**Mainly earthworms and grain; also insect
larvae, potatoes, peas, root crops, nuts,
snails, spiders, centipedes and bird eggs**

BREEDING
**Age at first breeding: 2 years; breeding
season: eggs laid late February to early May;
number of eggs: 3 to 5; incubation period:
16–18 days; fledging period: 32–33 days;
breeding interval: 1 year**

LIFE SPAN
Up to 20 years

HABITAT
**Open agricultural land, plains and steppes
with small woods and scattered trees; also
along coasts in winter**

DISTRIBUTION
**Europe except most of Scandinavia, east
through Russia and Turkey to Central Asia;
eastern Asia and southern Japan; introduced
to New Zealand**

STATUS
Common

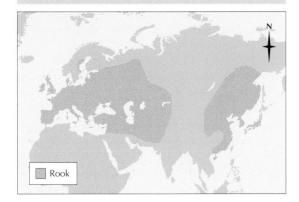

Rook

Sociable species

In the fall flocks of rooks take to the air and indulge in spectacular tumbling, diving and rolling displays. From a distance a flock looks like many large black leaves tossed about by the wind. Within the flock the rooks are more or less in pairs. At about this time they begin to show an interest in the nests, which are built in tall trees or sometimes in church towers. A clump of trees may have scores of nests in the upper branches, each built of sticks plastered with earth to consolidate them and lined with grasses, leaves, moss and roots. It is commonly said that the hen (female) builds the nest and the male brings her the materials. Close observation shows that the male alone builds it.

Breeding begins as early as late February, when three to five greenish eggs marked with gray and brown are laid. The hen incubates for 16–18 days, the male feeding her on the nest. He also feeds both hen and nestlings for the first 10 days after hatching, bringing food in his throat pouch. The rook's throat pouch, the opening to which is under the tongue, is more obvious than that of most birds because of the bare skin around the base of the bill. It bulges enormously when full and saves journeys in carrying food to the nestlings. It is also used in courtship, when the hen is fed ceremonially. The male flies over to his mate, his throat pouch bulging, and struts before her with his head held high, wings drooping and tail feathers fanned. Then he empties his throat pouch ceremoniously into hers.

Rooks probably have few predators because they readily combine to mob birds of prey. Humans are the rooks' worst enemy, and in places regular rook shoots are held to keep down their numbers.

*Rooks are a sociable
species of crow. They
nest in crowded treetop
colonies and in the fall
and winter they gather
in very large flocks.*

RORQUAL

THE RORQUALS NUMBER six species in all. They include the fin whale (*Balaenoptera physalus*), humpback whale (*Megaptera novaeangliae*) and blue whale (*B. musculus*), which formed the mainstay of the whaling industry for many years. They are baleen whales: each has plates of a bony substance in the mouth.

The largest rorqual is the blue whale, which is usually up to 72 feet (22 m) long, followed by the fin whale. The average length of the fin whale is about 60–65 feet (18–19.5 m). The sei whale, *B. borealis*, grows to 60 feet (18 m); Bryde's whale, *B. edeni*, grows to 50 feet (15 m); and the minke whale, *B. acutorostrata*, also called the lesser rorqual or pike whale, grows to about 33 feet (10 m). These six species form the family Balaenopteridae. All are broadly similar in appearance, with a large mouth and folds in the skin of the throat, a blue-black or pale gray back and paler whitish underparts.

Bryde's whale appears to be restricted to tropical and subtropical seas, whereas the other rorquals are found from North Polar to South Polar seas. Antarctic rorquals are generally larger than Arctic rorquals. Antarctic fin whales are about 8 feet (2.4 m) longer, on average, than those in the Arctic. This difference may be due to the more plentiful plankton in Antarctic waters.

Becoming rare

Rorquals live in groups of up to several hundred, although such large numbers are unlikely to be seen today because of overhunting. Since the boom in the Antarctic whaling industry during the 1930s, populations of most baleen whales have been severely depleted. The blue and humpback whales are now protected, and their numbers are recovering. After these two had been overexploited, whalers turned their attention to the fin whale and then to the sei whale. In the 1930s the Antarctic fin whale population was estimated at 250,000; by 1964 it was 50,000. Today it is estimated at about 120,000.

Apart from Bryde's whale, which stays in warmer waters, rorquals migrate from the polar regions in winter. While migrating they do not hug the coast as closely as the humpback whale, although the northern populations stay closer to the coast than those in the south. In winter the Antarctic population of fin whales appears to congregate in areas where food is plentiful, such as the coast of northwestern Africa, the Bay of Bengal and the Gulf of Aden. On their way south the part of the Arctic population that follows the European coastline detours to the west of the British Isles and rarely enters the North Sea. This is probably because the whales are following a route that existed long before the Straits of Dover opened, separating southeastern England from northwestern France.

Not all fin whales migrate toward the equator as winter approaches. Some stay put all year. In their summer quarters fin whales live outside the pack ice, but minke whales enter and are sometimes trapped. Sei whales spend the summer even farther from the pack ice zone.

MINKE WHALE

CLASS	**Mammalia**
ORDER	**Cetacea**
FAMILY	**Balaenopteridae**
GENUS AND SPECIES	**_Balaenoptera acutorostrata_**

ALTERNATIVE NAMES
Pike whale; little piked whale; lesser rorqual

WEIGHT
16½–132 tons (15–120 tonnes)

LENGTH
Head and body: 33–82 ft. (10–25 m)

DISTINCTIVE FEATURES
Relatively small, triangular head; 50 to 70 folds of skin on throat, allowing huge expansion of mouth cavity when feeding; dark gray upperparts; pale gray or white underside to belly and flippers; pale band on flippers, particularly in Northern Hemisphere

DIET
Crustaceans, including krill; occasionally fish and squid

BREEDING
Age at first breeding: 7–15 years, perhaps longer; breeding season: December–May (Northern Hemisphere); number of young: 1; gestation period: about 310 days; breeding interval: usually 2 years

LIFE SPAN
Up to 60 years

HABITAT
All oceans

DISTRIBUTION
Virtually worldwide

STATUS
One of the most common large whales, but many populations depleted due to whaling

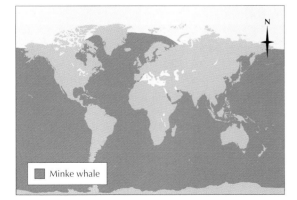

Minke whale

Sieving and gulping

The usual food of rorquals is planktonic crustaceans such as krill, copepods and amphipods, but fish and squid are occasionally eaten.

A rorqual feeds in the same way as any other baleen whale, straining its food through the baleen plates inside its mouth. There are two feeding methods: gulping and sieving. In the former the whale takes a mouthful of water and then shuts its mouth, squeezing the water through the baleen. The minke whale, Bryde's whale, fin whale, blue whale and humpback whale use this method. In contrast, right whales (discussed elsewhere) use the sieving method. They plow through the sea with their mouths open, letting the water rush through the baleen. When they have collected a mouthful of food, they shut the mouth and swallow. The sei whale sometimes uses the sieving method, which, where food is scarce, is more efficient than the gulping method.

Long-lived whales

Mating and birth occur in the winter, when the whales are in warm water, but the nonmigratory Bryde's whale mates and breeds year-round. Gestation lasts about 1 year and there is one young. Lactation lasts about 6 months and a cow breeds every other year. A calf's length at birth ranges from 9 feet (2.7 m) in the minke whale to 22 feet (6.6 m) in the fin whale. By the time it is weaned the calf has nearly doubled its length. Sei whales are sexually mature at 1½ years. Male fin whales are sexually mature at 5 years, but females take up to 8 years. Maximum size is reached at about 20 years, and fin whales are thought to live for up to 80 years.

Humpback whales are known for their unique songs. Scientists believe the songs are produced by solitary males in their tropical breeding grounds and are sung to attract mates.

ROSS SEAL

Ross seals are clumsy on the Antarctic pack ice but are strong and agile swimmers. They hunt mainly squid.

THE ROSS SEAL IS THE smallest of the various Antarctic seals, a little smaller than the crabeater seal, *Lobodon carcinophagus*. It probably averages 5½–8 feet (1.7–2.5 m) in length, although a male measuring about 9⅘ feet (3 m) has been recorded. The species weighs 290–440 pounds (130–200 kg), with a typical adult male weighing 390 pounds (177 kg). Newborn pups are 38–47 inches (0.96–1.2 m) in length and weigh about 36 pounds (16.5 kg), although few have been measured.

Plump, dumpy seal

The Ross seal has a markedly different appearance than that of other seals due to its wide head, short snout and small mouth. Its body is plump and its neck is short and thick, so its head seems to grow out of its shoulders. The muzzle is very blunt, which adds to the animal's stocky appearance, as does the seal's ability to withdraw its head into rolls of fat around the neck.

The Ross seal's coat is dark gray or tan above, paler underneath, with grayish streaks on the sides. A light and dark pattern around the eyes gives the head a masklike appearance. There are often broad dark stripes from chin to chest and on the sides of the head. Ross seal pups are dark brown above, paler beneath.

One of the Ross seal's most distinctive features is the large, protruding eyes, which in the past led to it being called the big-eyed seal. Some scientists believe the seal's large eyes have been adapted for hunting in dimly lit water, and that its sharp, needlelike teeth are ideal for catching slippery prey such as squid.

Rarely encountered by humans

The Ross seal is the least frequently observed of all the Antarctic seals, and close-up photographs of the species are few and far between. Almost all sightings of Ross seals have been made in the Antarctic pack ice. By 1945 only 50 had been recorded since British explorer James Ross, commander of HMS *Erebus*, found them in 1840. The distribution of sightings mostly reflects the spread of human activities in the Antarctic. The most northerly record is from Heard Island, north of the pack-ice limit.

Guessing their habits

Since the increase of human activity in the Antarctic region from the mid-20th century onward, Ross seals have been seen more often. They are, however, nearly always observed from the decks of ships or from helicopters and they are rarely, if ever, seen from the coastline of Antarctica or the neighboring islands. As a result, it has not been possible to study their habits to the extent that has proved to be the case with Weddell seals, *Leptonychotes weddelli*, which land on ice near the shores.

Rather than form breeding colonies, Ross seals are like the crabeater seal and the leopard seal, *Hydrurga leptonyx*, both of which also breed in widely scattered groups on the pack ice. Ross seals are quite clumsy on ice and react to the approach of humans by rearing up and gaping. At the same time they trill, a unique noise said to be produced by inflating the soft palate. Ross seals molt in late summer, during which time they do not eat. Little pieces of skin are shed during the molt, which recalls the molt of the elephant seals, genus *Mirounga*.

Squid diet

Their rather thickset bodies and specialized flippers suggest that Ross seals are strong, agile swimmers. This seems very likely since their main food appears to be quick-swimming squid, together with some fish and crustaceans such as krill. Ross seals catch their prey in the waters below the pack ice. Although their mouths are small the canine and incisor teeth have sharp, curved points suitable for holding slippery prey. The molars of the Ross seal are weak and are often lost early in life.

ROSS SEAL

CLASS	**Mammalia**
ORDER	**Pinnipedia**
FAMILY	**Phocidae**
GENUS AND SPECIES	***Ommatophoca rossi***

ALTERNATIVE NAME
Big-eyed seal

WEIGHT
290–440 lb. (130–200 kg)

LENGTH
Head and body: 5½–8 ft. (1.7–2.5 m)

DISTINCTIVE FEATURES
Smallest species of seal in Antarctic. Broad head; thick neck, tapering to slender body; very large, protruding eyes; proportionately small mouth; coarse coat; dark gray or tan upperparts; lighter underparts; specialized flippers.

DIET
Mainly squid; also fish and crustaceans such as krill

BREEDING
Age at first breeding: 3–4 years; breeding season: late December; gestation period: about 335 days, including 2–3 months of delayed implantation; number of young: 1; breeding interval: 1 year

LIFE SPAN
Probably up to 20 years

HABITAT
Antarctic pack ice and surrounding seas

DISTRIBUTION
Antarctic waters

STATUS
Little data available but not considered to be threatened; estimated population: at least 130,000

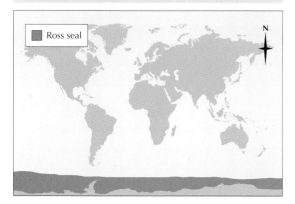

Ross seal

How rare are they?

In 1993 Dr. Peter Reijnders of the IBN (Institute for Forestry and Nature Research) suggested that the population of Ross seals was at least 130,000. Since scientists know so little about Ross seals' habits, this must remain a guess, but they are not thought to be a dwindling species. Ross seals are unlikely to be endangered by humans because of their remote habitat; even icebreakers steer clear of pack ice if they can. The main predator of Ross seals is likely to be the orca or killer whale, *Orcinus orca*. The seals have little chance of surviving against packs of killer whales unless they can get out of the water.

Surveys are now being done of the seals living in pack ice, observations being made either from ships or from helicopters. However, counts are difficult to interpret because the number of seals basking on the ice depends on the weather and probably on other circumstances as well. It has even been suggested that Ross seals live more in open water than in pack ice.

One survey of an area of over 700 square miles (1,800 sq km) found Ross seals to be nearly as numerous as leopard seals. In fact, 22 Ross seals were seen in 10 days, a remarkable increase over past records. Carleton Ray, the American zoologist who carried out the survey, suggests that Ross seals are often missed because they are easily mistaken for Weddell seals unless they rear up and gape in their characteristic aggressive posture. He also points out that because Ross seals are thought to be such a rarity, they are never recorded unless the observers are certain of the seals' identity.

The Ross seal has a highly distinctive habit of rearing up and gaping its mouth when danger threatens.

RUFF

Ruffs (male in winter plumage, above) feed in shallow water, along muddy shores and on damp grassland.

THE RUFF IS A SANDPIPER, the male of which has an ornate ruff of feathers around the neck in summer. The plain female, which has no ruff, is often called the reeve. Ruffs provide an extreme example of sexual dimorphism, a phenomenon in which the sexes are different in appearance.

Difference between the sexes

The male ruff is 10¼–12⅔ inches (26–32 cm) long, whereas the female, or reeve, is just 8–10 inches (20–25 cm) in length. The male is much stockier than the female, typically weighing 4⅔–9 ounces (132–255 g); the more slender female weighs 2½–6 ounces (70–170 g).

In summer the male ruff is readily distinguished from all other birds by his magnificent ruff and ear tufts. It is unique in the wide variation in plumage from one individual to another. There is an unlimited number of color combinations of the ruff, ear tufts and upperparts, including the tail and secondary flight feathers. There are, however, several main patterns with many variations between them. The upperparts may be black or brown with a purplish gloss,

white with black or brown freckles and bars or chestnut with black or brown markings. The ruff and ear tufts may be sandy, chestnut, purple or white, either uniformly colored or with bars and spots of black or brown.

Male ruffs in winter plumage resemble the females, and are sometimes difficult to distinguish from other shorebirds. They have little in the way of prominent features except for an oval patch of white on each side of the tail. Both sexes have scaly, dark brown upperparts and males have a sandy-brown crown and nape.

Ruffs breed in northern Europe and Asia from the Netherlands and Norway east to the Bering Strait. They were once quite common as breeding birds in the British Isles, but today only a few pairs breed in southeastern England.

Long migrations

Most shorebirds are found along the coast, on shores or in estuaries, but the ruff breeds inland and is less often seen near the sea. It prefers damp meadows, fens and marshes, from the damp surrounds of sewage farms to the boggy pools of the tundra. Outside the breeding season

RUFF

CLASS	**Aves**
ORDER	**Charadriiformes**
FAMILY	**Scolopacidae**
GENUS AND SPECIES	***Philomachus pugnax***

ALTERNATIVE NAME
Reeve (applies to female only)

WEIGHT
**Male: 4⅔–9 oz. (132–255 g);
female: 2½–6 oz. (70–170 g)**

LENGTH
**Head to tail: male, 10¼–12⅔ in. (26–32 cm);
female, 8–10 in. (20–25 cm)**

DISTINCTIVE FEATURESS
**Long legs. Breeding male: long ear tufts;
large ruff of feathers around head; variable
plumage, including white, brown, orange
and black. Female and nonbreeding male:
no ear tufts or ruff; gray and brown in color.**

DIET
**Mainly insects; also crustaceans, mollusks,
spiders, worms, small fish, frogs and seeds**

BREEDING
**Age at first breeding: 3 years (male),
2 years (female); breeding season: eggs laid
May–July; number of eggs: 4; incubation
period: 20–23 days; fledging period: 25–28
days; breeding interval: 1 year**

LIFE SPAN
Up to 11 years

HABITAT
**Summer: damp grassland near water. Rest of
year: coasts, marshes, lake shores, wet fields.**

DISTRIBUTION
**Summer: across northern Eurasia. Winter:
western and southern Europe; Africa; India.**

STATUS
Locally common

Ruff ☐ Breeding ▨ Nonbreeding

*Male ruffs display to
each other at traditional
display grounds known
as leks. Every male has
a different mix of colors
in his breeding plumage.*

ruffs gather in small flocks of about 20, but
several hundred may travel together on migration, when they cover vast distances to spend the
winter as far south as South Africa, India and Sri
Lanka. They feed and migrate mainly at night.

In winter ruffs are usually found on the
muddy margins of lakes, ponds, rivers and
marshes. They also frequent flooded fields,
particularly of wheat and rice. In certain areas
ruffs can be seen on very salty or alkaline waters.

The total population of ruffs is approxi-
mately 2 million birds, of which half winter in
West Africa.

Probing for food

Ruffs feed in shallow water or among damp
herbage, probing the mud or searching among
the plants for small animals. Their food is mainly
insects such as grasshoppers, bugs, beetles,
caddis flies and mayflies. They also eat snails,
worms, small crustaceans and spiders. In winter
especially, ruffs feed on the seeds of rice and
sorghum. Occasional food includes a variety of
small fish and frogs.

Sexes live apart

An unusual aspect of the ruff's behavior is that
the sexes tend to associate only during the
mating season. Even in winter males and females
usually stay in separate flocks, and the females
rear their broods alone.

The male ruffs gather on low hills, the equivalent of the prairie chickens' booming grounds.
They arrive in early April, before their elaborate
breeding plumage has fully appeared. Each ruff
stakes out his own station, which can be clearly

seen as a patch of ground about 2 feet (60 cm) across where the grass is flattened. The male ruffs gather on the hills, especially in early morning and evening, and display to each other. As each bird lands on his station, the others signal, flapping their wings to show their mainly white undersides.

Frantic displays

In the first stage of a display a pair of male ruffs runs frantically to and fro, with head and neck horizontal, ruff expanded and wings fluttering. Next the two males stop and crouch close to the ground with ruff, wings and tail spread, and they may quiver their wings. Displays are often set off by one ruff intruding into another's territory, and the trespasser is usually driven away by the displays of the owner. Fighting is not common, but when it does occur, it is quite violent, each ruff trying to get above its opponent and then beating it with its wings and pecking.

The females are attracted to the hills, and one female may visit several such display grounds. As a female alights, the males go into their crouching display. She may then walk away from the hill followed by some of the males,

A male ruff in the process of molting out of his breeding plumage. When the postbreeding molt is complete, it is time to migrate to the winter quarters.

but eventually she crouches in invitation and one mates with her. It has been found that a dominant male may take part in as many as 70 percent of the matings. Dominant males fight less than the others, presumably being assured of their status, but mate more often. Yet it is not the dominant ruffs that choose the females but the other way around. When a female has attracted the attention of a number of crouching males, she approaches one and nibbles the feathers of his head or ruff. Somehow the female can tell which is the best-quality male. When several females visit a display ground at once, they often stand in line for such a male.

Females lead families

The females nest close to the display hills, among grasses or rushes. The nests are simple hollows and are difficult to find unless the females are flushed out. Each female lays four eggs and incubates them for 20–23 days. The chicks leave the nests shortly after hatching, and for a few days their mothers catch insects for them. Later the young catch their own, but the females continue to guard them. Predators include gulls, jaegers and Arctic foxes, *Alopex lagopus.*

SABLE ANTELOPE

THE SABLE ANTELOPE, *Hippotragus niger*, and the roan antelope, *H. equinus*, are two of the most striking and most aggressive antelopes. They form a genus closely related to the oryx but differ in being larger. They have a short, high mane, and their horns, which are more upright than those of the oryx, are set on a slightly raised ridge just behind the eyes; the horns are also more strongly curved than those of the oryx. Sable antelope stand 51–57 inches (1.3–1.43 m) at the shoulder and bulls may weigh up to 595 pounds (270 kg). Roan antelope are taller and more long-legged, standing 56–60 inches (1.4–1.5 m) at the shoulder, but they are a similar weight to the sable antelope.

The male sable antelope is brown, black or nearly black. Females vary from golden brown to nearly black, the paler-colored females occurring in the north of the range. Both sexes have white lips and a white muzzle, and a white line from eye to muzzle. There are small tufts below the eyes, in the white face stripe. The upper part of the throat and the underparts are white; these areas stand out sharply against the dark color of the upperparts.

Roan antelope have a higher mane than sable antelope and a throat fringe. Their coloration is variable, although generally they are grayish roan in color with browner legs. The forehead and sides of the face are black or dark chestnut, and there is a long white tuft below each eye. The muzzle and underparts are white. In both species the young are light reddish. They can be distinguished by the more contrasting face marks of the sable and the shorter legs and ears.

In the sable antelope the ears are only half as long as the head; in the roan antelope they are tufted and nearly as long as the head. The horns of the sable are much longer than those of the roan, which are more flattened and more closely ringed and grow to a maximum of 38 inches (95 cm) long. The largest sable horns recorded were 70 inches (1.75 m) long.

Sable antelope are found from the Transvaal in the Republic of South Africa north to Angola and extreme southeastern Kenya. Roan antelope are more widespread, reaching western Ethiopia in northeastern Africa and Senegal in the west.

Water-hole priority

Sable antelope live in herds of 8 to 40, consisting of several females and young with a single black bull. Young bulls seem to leave the herd as they reach maturity and begin to turn black. Not all of them can form harems, and many are solitary.

They live in light bush country and graze in the early morning and the evening, lying up during the heat of the day. Every morning and evening the herd visits the water hole, drinks quickly and then leaves again rapidly. Other animals yield right of way to them and let them pass.

Roan antelope browse rather than graze and live in smaller groups than sable antelope, often in pairs. There are also solitary bulls, which may associate with herds of zebra or wildebeest.

Breed year-round

The herd bull fiercely protects his harem, and fights commonly break out whenever a solo bull approaches the herd. There is no restricted rutting season. Sable antelope have a birth peak that varies according to the region. In both species the gestation period is 268–286 days.

The coats of most male sable antelope become progressively more black as they mature. They also develop white patches on the face and belly.

Herds of sable antelope generally comprise females and young presided over by one bull. During the rutting season young males round up 10 to 20 females for breeding.

Humans the main danger

Both roan and sable antelope can be aggressive and dangerous. They and their relatives the oryx are among the few antelopes that will defend themselves vigorously when attacked. They kick and bite and make sideway sweeps with their curved horns. At one time sable were hunted with dogs, which usually resulted in a high mortality rate among the dogs. The antelopes are not intimidated by lions, and little else preys on them in the wild. When they are disturbed, sable run off in a bunch, roan in single file.

Dwindling numbers

Today the sable antelope is rather rare and sporadically distributed. The largest, longest-horned subspecies is the giant sable antelope of Angola, *H. niger variani*, which is one of the world's most endangered animals. The giant subspecies is restricted to a small area bounded by the Luando, Danda, Cuanza and Luasso Rivers in Angola. Discovered in 1916, there are only 500 to 700 today, of which about one-third are in the Luando Reserve.

The fauna of South Africa was soon depleted by the Boers, who took their ox wagons out on meat hunts among the big herds of game, leaving far more animals to die on the veld (open grassland with sparse shrubs or trees) than they collected for meat. The zebralike quagga, *Equus quagga*, vanished completely, along with several other distinct subspecies, while the black wildebeest (*Connochaetes gnou*), blesbok (*Damaliscus phillipsii*), bontebok (*D. dorcas*) and others survived only on enclosed land. The first South African animal to be exterminated was a member of the sable-roan group, the bluebuck, *H. leucophaeus*, which died out early in the 19th century.

SABLE AND ROAN ANTELOPES

CLASS **Mammalia**

ORDER **Artiodactyla**

FAMILY **Bovidae**

GENUS AND SPECIES **Sable antelope, *Hippotragus niger*; roan antelope, *H. equinus***

WEIGHT
***H. niger*: 440–595 lb. (200–270 kg).
H. equinus: 507–660 lb. (230–300 kg).**

LENGTH
***H. niger*, head and body: 84–102 in. (2.1–2.55 m); tail: 18–30 in. (45–75 cm).
H. equinus, head and body: 88–98 in. (2.2–2.45 m); tail: 24–28 in. (60–70 cm).**

DISTINCTIVE FEATURES
***H. niger*: brown-black upper body; brown mane; white face mask. *H. equinus*: short, smooth gray, brown or amber coat on upperparts; white to yellow undersides; gray or black neck, mane and tail. Both species: very large, backward-curved, ringed horns.**

DIET
Grasses and foliage of low shrubs

BREEDING
Age at first breeding: 2½–3 years; breeding season: varies according to rain; number of young: usually 1; gestation period: 268–286 days; breeding interval: 1 year

LIFE SPAN
Up to 17 years in captivity

HABITAT
Dry scrub savanna and forest edges

DISTRIBUTION
***H. niger*: eastern sub-Saharan Africa.
H. equinus: western sub-Saharan Africa.**

STATUS
Both species at low risk overall; subspecies *H. n. variani*: critically endangered

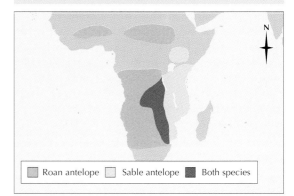

Roan antelope   Sable antelope   Both species

SAIGA

THE SAIGA AND ITS RELATIVE the chiru, *Pantholops hodgsoni*, are noted for their puffy noses, which probably warm the air they breathe and conserve water. The saiga, *Saiga tartarica*, is an antelope with some of the characteristics of the tribe Caprini, or sheep-goat group, and others of the tribe Antilopini, the gazelle group. Its relationships are obscure, and different scientists have classed it first with one group and then with the other. The saiga has long incisor teeth like those of sheep, but several of its skull characteristics and the glands on its forelegs, the feet, the face and in the groin are like those of gazelles. It stands 24–32 inches (60–81 cm) high and weighs up to 152 pounds (69 kg). Its horns, restricted to the male, are lyre-shaped, strongly ringed, about 12 inches (30 cm) long and a semi-transparent amber color. In summer its woolly upper coat is cinnamon in color with white underparts and tail; in winter its coat is white.

A smaller subspecies, the Mongolian saiga, *S. t. mongolica*, is only 24–27 inches (60–69 cm) high with thin, underdeveloped-looking horns and a dull grayish coat. The Tibetan antelope or chiru has been shuffled between the Caprini and Antilopini. Although it is the same size as the saiga, the chiru has extremely long horns, over 2 feet (60 cm) long. Its fur is very dense, woolly and pale fawn. The chiru is hunted and poached heavily for its wool. Its face is black and there is a black stripe down the front of each leg. It has glands in the groin and lateral hooves.

Both the saiga and the chiru have a sac inside each nostril, which is lined with mucous membranes that moisten and warm the inhaled air. This is probably very necessary in the cold, arid climates in which they live. Saiga are found across the Russian steppes from Kalmyckia, just west of the Volga River, into Dzungaria in Central Asia. The Mongolian saiga lives in the Gobi Desert and the chiru in the Tibetan Plateau, from Ladakh to Szechwan.

Massive migrations

The home of the saiga is the vast, lonely steppe, where there are no trees, only low-growing cespitose and wormwood, on which the saiga feeds. Huge herds constantly break up and re-form. After the rut in December, the harems and the rams come together, often in herds of over 1,000. In the early spring they disperse again, forming herds of usually 50 to 100. These herds migrate northward in search of new pastures, sometimes covering up to 200 miles. The males go first, followed by the females, forming herds of up to 100,000. In April the ewes form birthing grounds where they give birth, and most of the rams live alone or in small groups. Gradually the pastures deteriorate, the large concentrations of saiga split up and the irregular wandering becomes the rule once again. In August and September the saiga migrate south, gradually re-forming into big herds that do not break up again until the rut.

During the migration the herds move at 3–12 miles per hour (5–19 km/h). Sudden panics set them running, and the whole herd masses together and runs for miles at up to 50 miles per hour (80 km/h), their heads held low off the ground. At this time the mobile and down-pointing nostrils come into their own, keeping out the dust. While running, individuals take occasional leaps just to have a look around.

Fighting rams

In the December rut the rams fight for harems, some gathering as many as 50 ewes, while others have none and are forced to join the bachelor herds. The older females come into heat first, and there is a mass mating period that lasts about a week. During the last few days of the rut the younger females come into heat. About two-thirds of all births are twins. Of the rest, half are single births and half are triplets. For their first

The saiga has a prominent nose that warms the inhaled air. The female (below) lacks the lyre-shaped horns of the male.

The Tibetan antelope or chiru has extremely long horns and lives on the high Tibetan plateau, from Ladakh to Szechwan.

SAIGA

CLASS **Mammalia**

ORDER **Artiodactyla**

FAMILY **Bovidae**

GENUS AND SPECIES ***Saiga tartarica***

WEIGHT
57–152 lb. (26–69 kg)

LENGTH
Head and body: 39–55 in. (100–140 cm); shoulder height: 24–32 in. (61–81 cm); tail: 2½–4¾ in. (6–12 cm)

DISTINCTIVE FEATURES
Swollen nose; heavy, woolly, fringed, cinnamon-colored upper coat; dark cheeks and nose; white underparts and tail

DIET
Grasses, herbs and shrubs

BREEDING
Age at first breeding: 8 months (female), 2 years (male); breeding season: November–December; number of young: usually 1 or 2; gestation period: 7–8 months; breeding interval: 1 year

LIFE SPAN
Up to 10 years

HABITAT
Tundra and steppe

DISTRIBUTION
Mongolia and southern Russia

STATUS
Vulnerable

few days the young lie where they are born, hidden in cover. Afterward they keep close to their mothers. Births occur in April after a gestation period of 7–8 months. The mortality of lambs is very high: 10 percent are eaten by wolves, foxes, snow leopards, feral dogs or eagles or are simply lost by their mothers and starve before they even leave cover. The female defends them, however, if she is near, jumping into the air and lashing out with her hooves. Predation and starvation account for another 10 percent within a month. The cold, snowy winter kills many more, and by the following April only 40 percent of the lambs remain.

Female saiga become sexually mature at 8 months, and 85 percent of young females become pregnant, compared with 96 percent of females more than a year old. Males are not sexually mature until they are about 2 years old. At all ages the mortality among males is much greater than that among females. A female's life expectancy is up to 10 years, but a male's is only 5–7 years. Many males are so exhausted by their rutting fights that they fall easy prey to wolves, if they have not already killed each other.

Chiru, which live in Tibet at 12,000–18,000 feet (3,660–5,490 m) altitude, also rut in early winter. Their battles are equally fierce and cause a high mortality among the rams. The young are not born until June or July. The female scrapes out a shallow pit in the ground and deposits her young there.

Chiru consistently lie huddled in depressions in the ground. When they spot danger, they rise together and run with their tails up and heads down, but soon stop to look back. Like saiga, they are constantly on the move in search of good pasture. They move into swampy valleys when the snow melts to feed on fresh grass, and during the summer months they migrate to the

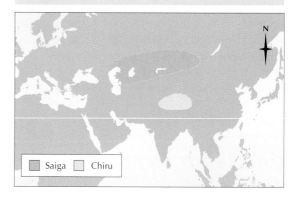

highest valleys. Overhunting, droughts and severe winters in recent years have all combined to reduce drastically the numbers of saiga, but under strict protection, their numbers have slowly increased, providing meat and hides. However, they remain vulnerable.

SAILFISH

THIS IS A HIGHLY streamlined fish that can swim remarkably fast. Its upper jaw is prolonged into a swordlike bill, and the first dorsal fin is large and forms a sail when fully raised. The second dorsal fin and the two anal fins are small. The narrow, sabrelike pectoral fin is also small, and the pelvic fins, set forward under the throat, are longer, narrower and even more sabrelike. Keels on each side lie just in front of the strongly crescent-shaped tail fin.

Sailfish are classed in two species: *Istiophorus albicans,* the Atlantic sailfish, and *I. platypterus,* living in the tropical waters of the Indian and Pacific Oceans. In warm summers, Atlantic sailfish venture as far north as Cape Cod and the Gulf of Maine. Although widespread, the Atlantic sailfish is now under pressure from overfishing in the Caribbean and off Florida.

Built for speed

The outstanding feature of the sailfish is its speed. It is one of the fastest fish, perhaps the fastest of all. Tests in the 1920s measured the length of fishing line pulled by a sailfish. In 3 seconds, a sailfish reeled out 100 yards (90 m) of line, which equates to an incredible speed of 60 knots (111 km/h). Conservative writers now prefer to quote a speed of 20–30 knots (37–55 km/h). When swimming at speed, the sail is folded down to lie in a groove along the back. The long pelvic fins are drawn up under the body, and the pectorals lie flush with the sides.

Sailfish live in the turbulent, oxygenated waters of the oceans. They have sievelike gills, which have a large surface for taking in oxygen, a necessity when traveling at great speed. It appears that very high speeds are used only in short bursts and that the conservative 20–30 knots may refer to its cruising speed. It seems also that the sail is erected, as it is in most fish, at the end of a burst of speed to prevent the body from rolling and yawing.

The vertebral column is especially strong; the vertebrae are tightly interlocked by horizontal processes, and the dorsal and neural spines, the long, thin processes that extend from each vertebra, are flattened, forming a strong and rigid backbone. These flattened surfaces also provide anchorage for the powerful muscles that drive the fish through the water.

Small fish put to the sword

The upper jaw of the sailfish has many small teeth, and these extend forward onto the lower surface of the swordlike bill. Sailfish eat other fish, especially flying fish. They also eat the deep water false albacore, the snailfish (genus *Liparis*), and the gurnardlike sea robins (family Triglidae) that live on the bottom. They may catch needlefish and anchovies. Sailfish also eat squid and octopuses, although how they catch them is a matter for debate. Some say they beat them with their spearlike bills, whereas others maintain that they snap them up with their teeth. They probably do both, but when fish are used as bait, they are always bitten on both sides by the sailfish. When a sailfish attacks a shoal of small fish, it swims around the shoal with its dorsal fin half-raised, driving the fish into a compact mass. Then it swims through this mass, thrashing vigorously from side to side with its bill, killing or stunning large numbers. The sailfish then swims around slowly, picking up the fish as they sink.

Those that get away

The sailfish gives exciting sport for the sea angler, as it leaps out of the water and thrashes around, putting up a spectacular fight. It is also good to eat, but sport anglers and conservationists now discourage the use of sailfish in restaurant menus. Sport fishing is strictly regulated. Scientific study and conservation are

The sail-like dorsal fin of the sailfish might look cumbersome, but when the fish travels at high speed the sail can be laid flat in a groove on the back.

Juvenile sailfish (preserved specimen, above) develop their large, sail-like dorsal fin and elongated jaws when only 2 inches (5 cm) long.

both linked with the sport. In Florida, a large proportion of hooked sailfish are not pulled aboard but tagged and set free. To encourage the liberation of fish, fishing and conservation clubs award extra points in competitions for specimens released. Above all, this helps in monitoring sailfish numbers and alerts scientists to population declines. However, by tagging, scientists also gain vital information on the development, longevity and migration of sailfish. They turn out to be highly migratory, tagged specimens being recovered up to 1,900 miles (3,000 km) away from where they were first caught.

Fragile babies

When spawning occurs, males and females swim in pairs, or two or three males may chase a single female, probably a mating behavior. The eggs are shed by the female in great numbers. From spawning to maturity, the life of a young sailfish is a mystery, which is of great concern, because without knowledge of the species' life history, it is difficult to manage sustainable fishing. In recent years, scientists have tried to net and study baby sailfish. The larvae are caught in narrow-mesh plankton nets dragged for short distances by boat, then they are placed in buckets of water on deck. The aim is to maintain the baby sailfish in captivity, to study their feeding, behavioral and habitat needs. However, the tiny sailfish are impossibly fragile. If they survive the netting procedure, they tend to react badly to being confined in a bucket and repeatedly nose dive to the bottom. Maybe this is not surprising in a small, pelagic animal that has never encountered a solid surface. The baby sailfish survive only a few days in captivity at present, but scientists are confident they can relax baby sailfish in the future and accustom them to captivity.

SAILFISH

CLASS **Osteichthyes**

ORDER **Perciformes**

FAMILY **Istiophoridae**

GENUS AND SPECIES **Indian and Pacific Ocean sailfish, *Istiophorus platypterus* (details below); Atlantic sailfish, *I. albicans***

WEIGHT
Up to 220 lb. (100 kg)

LENGTH
Up to 11½ ft. (3.4 m)

DISTINCTIVE FEATURES
Slender body; high, sail-like first dorsal fin; pelvic fins long and slender; jaws elongated into a narrow bill; body dark blue above, fading to white or silver below; often dusky vertical bars or spots on flanks; dorsal fin bright cobalt blue with black spots

DIET
Fish, crustaceans and squid

BREEDING
Age at first breeding: 3 years; breeding season: spawns all year round, with peak in summer months; number of eggs: several million

LIFE SPAN
Up to 10 years

HABITAT
Oceans, usually in warm surface waters at 70–82° F (21–28° C); often near coasts and islands

DISTRIBUTION
West Pacific: from 45–50° N to 35° S; East Pacific: from 35° N to 35° S; Indian Ocean: south to 35–45° S; also in eastern Mediterranean and Red Sea

STATUS
Not threatened, but overfished locally

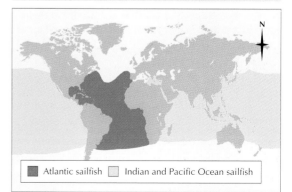

■ Atlantic sailfish ☐ Indian and Pacific Ocean sailfish

SAKI

SAKIS ARE SOUTH AMERICAN monkeys with striking, coarse coats of fur. There are two groups: the sakis, genus *Pithecia*, and the bearded sakis, genus *Chiroptes*. They all have bushy tails and thick dark fur as well as elaborate hair patterns on the head. In common with all New World monkeys, they have broad noses with nostrils pointing sideways, separated by a wide septum. Sakis have nails on their fingers and toes, though these are convex and rather pointed, almost clawlike. Their lower incisors are long and lean forward, a feature shared only with the related uakaris.

The male white-faced saki, *P. pithecia*, has a black coat, and a hood of hair originating in a whorl on the nape of the neck hangs over the forehead. The female's coat is more brindled (bearing dark streaks over a tawny or gray background) than that of the male. The sexes are highly distinctive: the male's face is white, whereas the female's is dark with white streaks running from the sides of the nose to the corners of the mouth. It lives in lowland and montane forests. The red-bearded or monk saki, *P. monachus*, has a grizzled gray-brown coat, with paler feet and hands. Its beard and underparts are reddish. It is found in lowland forests.

Sakis grow up to 40 inches (1 m) long, including the tail; the male is slightly larger than the female. They generally favor forest on high ground, although they are also found on savanna and in forest that has been disturbed. They are usually found in the lower and middle canopy levels of trees and are thought to be monogamous or to live in extended family groups consisting of two to five individuals.

The bearded sakis are larger than other sakis and have less distinctive faces. They have extravagant hairstyles, although their fur is of medium length, shorter than that of the sakis. The white-nosed saki, *C. albinasus*, is black with a pinkish white nose and upper lip. Its hair lies forward in a fringe over the forehead. It is found both on high ground and in swampy forests. The black or bearded saki, *C. satanas*, has a parting on the crown, on either side of which the hair rises in a bouffant style, and it has a bushy beard that is about as long again as the face. Like the white-nosed saki, it is entirely black apart from a light yellow-brown to dark brown color to the back and shoulders. It has swollen temples that are covered in fur. Bearded sakis occur exclusively in high-ground mature forests. They live in the upper tree canopy, although they are occasionally found in the understory. Bearded sakis are probably polygamous (have more than one sexual partner), living in groups of 8 to 30 individuals that include more than one male.

Most sakis and bearded sakis occur from the Orinoco River north of the Amazon to the Guianas and around the upper Amazon. The exception to this is the white-nosed saki, which is restricted to an area south of the Amazon between the Madeira and Xingu Rivers.

Spongelike fur

Sakis are active tree-dwelling monkeys. They move around by day, feeding mainly on fruits, including immature seeds, and certain leaves. Although spending most of their time in the high canopy, they descend at times to the shrub layer to feed. They drink by soaking the fur on the backs of their hands in water and licking it.

Many sakis are strikingly colored. The male white-faced saki (below) is told from the female by his white mask.

Like many of their relatives, sakis (female P. pithecia *pictured) grip objects between the second and third fingers; the thumb is not opposable, as it is in Old World monkeys and apes.*

SAKIS

CLASS	**Mammalia**
ORDER	**Primates**
FAMILY	**Cebidae**

GENUS AND SPECIES   **Sakis: white-faced saki,** *Pithecia pithecia*; **monk saki,** *P. monachus*; **bald-faced saki,** *P. irrorata*; **equatorial saki,** *P. aequatorialis*; **white saki,** *P. albicans*. **Bearded sakis: white-nosed saki,** *Chiropotes albinasus*; **black saki,** *C. satanas*.

WEIGHT
1½–7⁷⁄₁₀ lb. (700–3,500 g)

LENGTH
Head and body: 12–28 in. (30–70 cm); tail: 10–22 in. (25–55 cm)

DISTINCTIVE FEATURES
Bushy, thick hair; coat color varies between dark red, black, brown or tan according to species; facial hair contrasts with body hair in some species, such as *P. pithecia*

DIET
Fruits, leaves, flowers and honey; insects; small animals, including mice, bats and birds

BREEDING
Age at first breeding: 2½–4 years; breeding season: may occur year-round; number of young: 1; gestation period: about 165 days; breeding interval: 1 year

LIFE SPAN
Up to 35 years in captivity

HABITAT
Forests; bearded sakis at higher altitudes

DISTRIBUTION
Northern and central South America

STATUS
***P. m. milleri*: vulnerable. *C. albinasus*: endangered; probably rarest of all South American primates. Some others common.**

Sakis and bearded sakis (all species)

In captivity, however, they learn to drink with their lips. Naturalists believe that sakis tend to live more toward the forest borders, but bearded sakis reportedly live deeper in the forest, especially alongside rivers where the forest grows very thick as a result of the greater amount of daylight reaching the ground vegetation.

Little is known about the general behavior of sakis. They probably live in pairs and defend their territories like titis and marmosets. They jump and run through the trees, sometimes hanging by their arms. The young at first cling to the mother's belly. This is unusual: most New World monkeys sit on their mothers' backs.

To date no *Pithecia* species have been classed as endangered. However, all species are hunted for food and for the pet trade, and are potentially at risk. One subspecies of the monk's saki, *P. m. milleri*, is currently classed as vulnerable. The two bearded saki species are thought to be at more of a risk than other sakis, and the white-nosed saki is already classed as endangered. It is probably the rarest South American primate.

SALP

SALPS ARE BARREL-SHAPED, transparent animals living in the plankton. They are quite simple in form and use a remarkable method of reproduction. They are members of the tunicate group, Urochordata, which also contains sea squirts. The tunicates are thought to be a link between vertebrates and invertebrates. Salps have the same fundamental body plan as sea squirts, but while the sea squirt's inhalent opening, or mouth, is near the exhalent opening, or vent, both near the top of the body, the openings of salps oppose each other at either end of its body. Water is drawn in through the opening at one end, powered by cilia (hairlike, beating organs) inside the salp, and driven out at the other end. The water current brings in food and oxygen and creates a feeble form of jet propulsion that the salp uses for getting around.

Around a salp's transparent, jellylike body, bands of muscle form irregular or partial hoops that are somewhat similar to the hoops of a barrel. By the alternate contraction and relaxation of these bands the body pulsates like a heart, aiding the creation of the current it uses for feeding and locomotion.

Salps are found near the surface, but some migrate vertically to deeper waters every morning, rising again to the surface waters every evening. They are usually found at considerable distances from land, especially in the warmer waters of the world.

Curtain feeding

The internal organs of a salp are simple. The gills are little better than large openings in the gullet, the digestive tract little more than a simple tube, and the nervous system merely a knot of nerve cells with slender nerve fibers leading away from it. There are no special sense organs. A salp feeds on small particles in the water, which are trapped in a curtain of mucus, like an irregular spiderweb, that is continually drawn up from the floor of the gullet.

Alternation of generations

Salps have a complex life cycle in which they alternate between solitary, asexual forms and sexual forms found in aggregations of several linked individuals. They are therefore said to show alternation of generations. The sexual salps are hermaphroditic, carrying both eggs and sperms, but since these do not ripen at the same time, a salp is not self-fertile and its eggs must be fertilized by sperm from another salp. There are only one to three eggs at a time, connected to the mother by a kind of placenta through which they are nourished. When these eggs have developed into salps, they break away from the mother and her aggregation and become solitary. The solitary salp grows a tail-like process on its underside. A chain of new individuals buds off asexually from the tail. These individuals form the aggregated, sexual generation.

Salps are relatives of sea squirts, and swim freely. Like this Salpa *species pictured near Hawaii, they are usually found among the plankton in surface waters.*

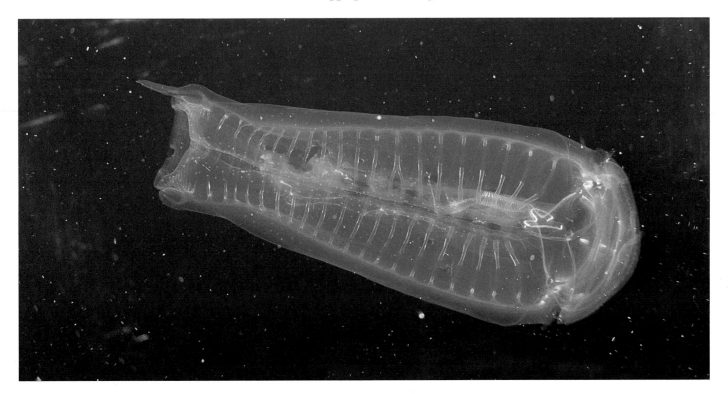

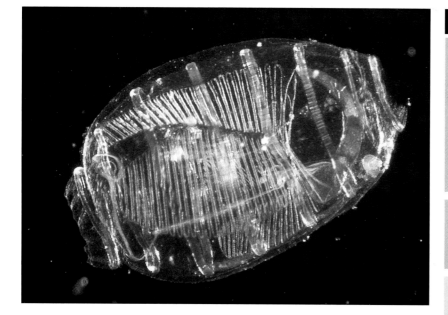

Doliolum is a pelagic animal that belongs to the same subphylum as salps. The sexual form, or gonozooid, pictured is just one stage in the highly complicated Doliolum *life cycle.*

SALPS AND RELATIVES

PHYLUM	**Chordata**
SUBPHYLUM	**Urochordata**
CLASS	**Thaliacea**
ORDER (1)	**Salpida**
GENUS AND SPECIES	**Giant salp,** *Salpa maximus* **(described below); many others**

ORDER (2)	**Doliolida**
GENUS AND SPECIES	*Doliolum nationalis*; **others**

LENGTH
Usually less than 2 in. (5 cm)

DISTINCTIVE FEATURES
Hollow, jellylike, barrel-shaped body; muscle bands in hoops around body

DIET
Microscopic particles filtered from water

BREEDING
Alternates between sexual and asexual generations; sexual reproduction hermaphroditic, with cross-fertilization essential; asexual reproduction by budding

LIFE SPAN
Probably less than 1 year

HABITAT
Warm, open waters, mainly at surface

DISTRIBUTION
Subtropical and tropical seas and oceans

STATUS
Abundant

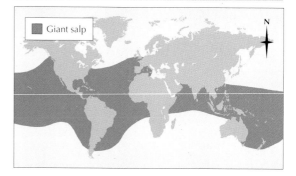

Buds without end

Doliolum is another barrel-shaped, jellylike animal. It is a relative of the salps, but is placed in a different order. Its muscle bands are more regular, and its way of life is similar to that of the salps except for its method of asexual reproduction, which is remarkably complicated. It begins with buds appearing on the underside of the barrel. The buds then creep up the sides of the barrel in a continuous procession as they are created, and make their way to a small, finlike tail at the rear of the parent body. The stream of buds settles on this tail in three rows extending along the top and side surfaces. Each bud in the top row grows a stalk, and new buds settle on these stalks. The buds in the side row start to feed, and they supply food for all the other buds as well as the parent body, which now becomes no more than a vehicle towing a chain of multitudinous progeny.

Occasionally, the buds in the top row break away and start new aggregations of buds. As these new aggregations continue to bud asexually, the progeny of this phase of budding begin to break loose, swim away and become sexually mature forms called gonozooids, producing ova and sperms. They form the sexual generation. The ova and sperms are later shed into the sea, and from each fertilized egg hatches a tadpole-like larva that turns into an asexual barrel, and the cycle begins all over again. It is not surprising that in places the sea is crowded with *Doliolum*.

Shrimp's baby carriage

A shrimp named *Phronima* preys on the salp. *Phronima* is 1 inch (2.5 cm) or so long with large compound eyes and exceptionally big claws. It enters the barrel and eats all the internal organs while using the barrel as a kind of luxury coach in which to travel. The female shrimp, having scraped out the transparent barrel, then lays her eggs in it. Then she pushes this adopted brood chamber in front of her, rather like a baby carriage, until the eggs are ready to hatch out and the larvae eventually swim away.

SAND DOLLAR

SAND DOLLARS, also known as cake urchins or sea biscuits, are closely related to sea urchins. Although flat rather than spherical, their shells, or tests, show the basic five-armed body pattern of echinoderms such as starfish, brittlestars and basket stars. Compared with these other echinoderms, sand dollars are bilaterally symmetrical, although in the center of the upper surface of the test there is a "star" of pinholes. A flat test is not as strong a shape as the globe tests of sea urchins, and the tests of sand dollars are strengthened inside by supporting pillars between the upper and lower surfaces. The test is usually 3–4 inches (7.5–10 cm) across and is purple or black in color.

The intact sand dollar is covered with short spines, about 1.5 millimeters long, which give it the appearance and texture of velvet, very different from the long, sharp spines of sea urchins. Unlike sea urchins and starfish, there is a noticeable forward end to a sand dollar, and the anus is located at the rear edge instead of at the center of the upper side.

Sand dollars are found mainly in the warm waters of the world. Most species live on the shores of the United States, South America and Japan, although the common sand dollar, *Echinarachnius parma*, is found as far north as Alaska and Canada's east coast.

Edging into sand

Sand dollars are found on sandy shores and tidal flats where they are not exposed to surf. Only in the quietest locations can they be found above the low tidemark, hiding themselves under the sand. Sand dollars are often very abundant and as many as 468 have been found in 1 square yard (0.8 sq m). They dig themselves just under the surface of the sand by plowing down at a shallow angle, propelled by the combined movement of the spines. Sand gradually accumulates in front of them, and over time the sand dollars become covered. Some species wholly submerge, although their outline can be made out, whereas others leave the rear end exposed. When covered with water, sand dollars stand on edge, with one-third of the body buried. They orient themselves so that the body is held at right angles to the flow of water.

Continuous feeding

The spines of a sand dollar are covered with cilia (short, hairlike projections) that beat, setting up minute eddy currents that draw in tiny organisms and particles from the water or sand surrounding the sand dollar. These organisms are trapped in mucus secreted by the spines. The mucus flows down the spines and is gradually passed down branched pathways that lie on the surface of the sand dollar. As the organism moves, this mucus accumulates into five large tracts that lead around the edge of the test and along the underside to the mouth at the center. Close examination of the test reveals that these tracts are fine grooves with even finer branches for filtering out unwanted material or organisms.

This selective method of feeding differs from that of many sand-dwelling animals, which eat, and eject, large amounts of sand and extract edible material as it passes through the gut.

Breeding

Sand dollars release eggs and sperms into the sea, where fertilization takes place. The small, delicate larvae resemble those of the acorn worm.

The waterworn test (shell) is the only part of the sand dollar that is usually visible, as the animal normally lives buried in sand or mud below the low tidemark.

The shells of some sand dollars, including *Mellita sexies perforata* (above), have several characteristic slits around the edges. This has given rise to the common name of keyhole urchin.

SAND DOLLARS

PHYLUM **Echinodermata**

CLASS **Echinoidea**

ORDER **Clypeasteroidea**

FAMILY **Clypeasteridae**

GENUS AND SPECIES **Several, including sea biscuit, *Clypeaster rosaceus* (detailed below); common sand dollar, *Echinarachnius parma*; eccentric sand dollar, *Dendraster exentricus*; and pea urchin, *Echinocyamus pusillus***

LENGTH
3–4 in. (7.5–10 cm)

DISTINCTIVE FEATURES
Flattened oval test (shell); 5 large petal-like ambulacra (radial areas) on upper surface; reddish, yellowish or greenish color

DIET
Waterborne detritus of all kinds

BREEDING
Sexes separate, but male and female rarely take on different forms; groups aggregate to spawn; male releases pheromones with sperm to stimulate egg release by female; free-swimming ciliated larvae develop into small sand dollars and settle on seabed after 3–6 weeks

LIFE SPAN
Possibly 5–10 years

HABITAT
Sand substrates or reef terraces at depths of up to 16½ ft. (5 m); sheltered shores

DISTRIBUTION
North Pacific: Alaska south to Washington State. North Atlantic: Newfoundland south to North Carolina.

STATUS
Common

After swimming freely for 3–6 weeks, the larvae undergo radical changes and settle on the seabed as young sand dollars.

Sheltering from danger

As well as a dense covering of spines, sand dollars have many tube feet and pedicellariae (slender parts on the base). One function of the pedicellariae is to prevent animals such as barnacles from settling on the sand dollars, and they may even secrete a poison. However, the pedicellariae and spines appear to offer little protection against marine snails and starfish, the former being able to rasp holes in the test with a radula (horny band that bears minute teeth on its dorsal surface). Accordingly, the sand dollars rely on burrowing for safety when faced with such predators. When a starfish passes through a bed of sand dollars, it leaves a clear track about 4 feet (1.2 m) wide where the sand dollars have hastily buried themselves, a process that takes 1–3 minutes. Starfish are slow-moving animals, and this is usually enough time for the sand dollars to submerge themselves.

The appearance of a starfish that does not eat sand dollars causes no alarm. It seems likely that sand dollars can recognize dangerous starfish, probably by chemical emissions from the bodies of such predators. Sand dollars are not the only animals with this ability. The Antarctic limpet, *Patinigra antarctica*, for example, flees from limpet-eating starfish but ignores other species.

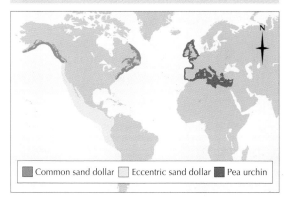

Common sand dollar   Eccentric sand dollar   Pea urchin

SAND EEL

SAND EELS ARE SMALL, eel-like fish of great economic importance. Their long bodies are either scaleless or have only very small scales, with none on the head. The lower jaw juts out beyond the upper jaw and is spoon shaped at its outer end. The upper jaw can be protruded to a great extent in some species. The pectoral fins are small, and of the 16 species of sand eels, only *Bleekeria mitonkurii* has pelvic fins. The dorsal fin is long and low, extending from just behind the head almost to the tail. The anal fin is similar but only half as long; the tail fin is forked. Sand eels are mostly 9 inches (23 cm) long but a few species reach 12 inches (30 cm). They are brown, blue or green on the back, usually with silvery flanks and a silver belly.

Sand eels are found mainly in temperate and cold seas in the Northern Hemisphere but some species are subtropical. The name sand lance is sometimes used for one or more of the species. Somewhat confusingly, the same name is applied to distantly related fish of the family Kraemeriidae, also called sand lances, which live buried in the sand in lagoons and streams at the edges of atolls in the Indian Ocean and South Pacific. Sand eels are widespread in the Arctic, Atlantic and Pacific Oceans.

Vast shoals

Around some coasts in summer when the seas are calm, vast numbers of sand eels are visible swimming about 6 feet (1.8 m) below the surface. If the water is suddenly disturbed, they all dive temporarily out of sight into the sand.

Of the 16 species of sand eels, 15 are from the family Ammodytidae and one is from the family Hypoptychidae. Each species lives at a different depth. For example, the smooth sand eel, *Gymnammodytes semisquamatus*, lives at depths of 60–600 feet (18–180 m); the small sand eel, *Ammodytes tobianus*, may be found at mid-tide level, where it can be dug up when the tide is out, but it also lives down to 90 feet (27 m). It lives on sandy or gravel bottoms, spending as much time buried in the sand as it does swimming. When entering sand or gravel, the fish pushes its spoon-shaped lower jaw among the grains by wriggling movements of the body.

Feeding

Sand eels eat crustaceans, including copepods, euphausians, amphipods and the larvae of crabs and lobsters. They also eat worms, eggs and the larvae of other fish, including other sand eel species. The young eat mostly crustacean larvae,

Huge shoals of sand eels may be found near the surface of the sea both inshore, from intertidal to subtidal areas, and offshore over deep water.

PACIFIC SAND EEL

CLASS	**Osteichthyes**
ORDER	**Perciformes**
FAMILY	**Ammodytidae**
GENUS AND SPECIES	*Ammodytes hexapterus*

ALTERNATIVE NAME
Pacific sand lance

LENGTH
Up to 10⅗ in. (27 cm)

DISTINCTIVE FEATURES
Elongated, eel-like body; tiny scales, with none on head; extensible upper jaw shorter than lower; very long, low dorsal fin, set in groove along back; forked tail fin; no pelvic fins; silvery flanks and belly

DIET
Zooplankton (planktonic animals)

BREEDING
Age at first breeding: 1 year; number of eggs: 10,000 to 30,000; hatching period: 14–21 days; breeding interval: about 1 year

LIFE SPAN
Up to 5 years

HABITAT
Near surface of both inshore and offshore waters; also buries in sandy seabeds

DISTRIBUTION
Northern Pacific Ocean

STATUS
Abundant

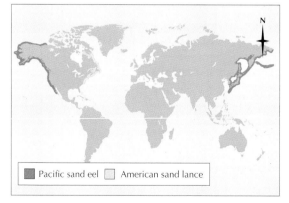

Pacific sand eel American sand lance

Many seabirds, such as the Atlantic puffin, Fratercula arctica (above), terns and guillemots, feed freely on sand eels, which are also a very important source of food for fish.

worms and fish eggs and larvae. This diet, which applies to the family as a whole, varies widely between species.

Buried eggs

The female sand eel, which is slightly larger than the male, lays 10,000 to 30,000 eggs. She bores her way down through the sand, laying her irregularly oval eggs as she goes. The eggs have a sticky surface and become coated with sand grains, which keeps them just slightly buried. They hatch 2–3 weeks later, the larvae being up to ¼ inch (6 mm) long. The larvae stay near the bottom for a while, swimming in very large numbers in the larger rock pools. They are then ⅖ inch (1 cm) long, and at this stage move up into midwater, where they grow rapidly. The adult small sand eel, for example, which is 8 inches (20 cm) long, reaches 4½ inches (11.5 cm) at the end of its first year. The great sand eel, *Hyperoplus lanceolatus*, which is 12½ inches (32 cm) long when adult, attains 5½ inches (14 cm) within the same period. After this the growth rate slows down. The life span of a small sand eel is 4 years; the great sand eel lives for about a year longer.

Sand eels are extremely common. In one year last century the German catch alone was over 13,000 tons (11,700 tonnes). Sand eels are caught in seine nets, trawls and shrimp nets; they are

also dug out and used as bait. Their chief role in the economy of the sea is in providing food for many other animals such as cod, haddock, halibut, herring, coalfish, terns, guillemots (murres) and puffins. Sand eels give rise to huge fisheries during summer and fall.

SAND GROUSE

IN SPITE OF THEIR name, sand grouse are more closely related to pigeons (order Columbiformes) than they are to the game birds of the order Galliformes, which includes the true grouse, although some writers think they are more closely related to waders and gulls (order Charadriiformes). Sand grouse are the size of small pigeons, being 9–16 inches (23–41 cm) long, and have very short legs with feathers right down to the toes, which may be partly webbed.

Sand grouse plumage is dense and prettily patterned, with a gray or brown background marked with black, white and orange. This makes the birds almost invisible in their desert or steppe homes. Sand grouse that live in open deserts are paler than those that live in places where the scant vegetation gives some shade. In the Kalahari Desert, sand grouse living on sand dunes have reddish plumage, whereas those living on outcrops of calcrete, a hard gray rock, have a matching gray coloration. The dense plumage with a thick underdown seems to give good protection from the sun, against which there is little other defense in these areas. Sand grouse walk awkwardly, but the flight feathers are long, and the birds are strong fliers. The 16 species of sand grouse live in dry parts of mainland Africa and Madagascar, Central and southern Asia and southern Europe.

Clockwork drinkers

Sand grouse live in flocks and fly to water in line abreast as regularly as clockwork. They may fly up to 40 miles (64 km) to a suitable pool and arrive at almost the same time each day, slightly earlier in winter than in summer. Sand grouse usually come to drink in the morning, soon after dawn, but some visit water in the evening. In the breeding season, both a morning and an evening drinking trip take place as sitting birds are relieved of incubation duties by their mates. Sand grouse favor certain pools or stretches of water, and in some places hundreds of thousands of birds may gather. They each take in as much as 1¼ ounces (37 ml) of water, enough to see them through a hot day. Sand grouse are particularly vulnerable when drinking and are wary as they approach water, especially in the evening, as this is the time when predators gather at water holes.

Adult sand grouse feed exclusively on hard seeds of annual plants or of trees such as acacias (genus *Acacia*). Some sand grouse chicks are also fed on seeds, and a chick just one week old was found with 1,400 small seeds in its crop. While some sand grouse are sedentary, others migrate regularly and there are occasionally massive migrations, like the irruptions of crossbills (genus *Loxia*), discussed elsewhere. Pallas's sand grouse, *Syrrhaptes paradoxus*, of Asia once irrupted eastward to Beijing and on several occasions invaded Europe as far as the British Isles, where migrants stayed to breed. In 1888 birds moved across Russia in March, with the advance guard reaching the North Sea in late April. Some easterly return movement was observed in late summer and autumn in mainland Europe but on a smaller scale than the initial irruption. Since 1908 irruptions have been rarer, possibly as a result of a contraction of the bird's west Siberian range.

Hidden nests

Sand grouse nests are no more than depressions in the ground, and some sand grouse often lay their eggs in the footprints of large animals such as camels. The nests are very difficult to find. In the course of a 19-month study, ornithologist G. L. Maclean found only one nest of a spotted

The Namaqua sand grouse (this is a male) is an inhabitant of southern Africa. It is one of the species known to transport water in its feathers to chicks at the nest.

The double-banded sand grouse, Pterocles bicinctus, *is a common resident of dry bushveld in southern Africa. This male was photographed in Kruger National Park, South Africa.*

PALLAS'S SAND GROUSE

CLASS	**Aves**
ORDER	**Pteroclidiformes**
FAMILY	**Pteroclididae**
GENUS AND SPECIES	***Syrrhaptes paradoxus***

WEIGHT
8½–10½ oz. (235–300 g)

LENGTH
Head to tail: 12–16 in. (30–41 cm)

DISTINCTIVE FEATURES
Orange-buff face; gray upper breast, buff lower breast and black belly; sandy-buff upperparts finely barred with black markings; white underwings; very long central tail feathers

DIET
Mainly seeds

BREEDING
Age at first breeding: 1 year; breeding season: eggs laid late March–July; number of eggs: 2 to 4; incubation period: 28 days; fledging period: 24–28 days; breeding interval: 2 or 3 broods per year

LIFE SPAN
Not known

HABITAT
Arid, open plains and uplands

DISTRIBUTION
Caspian Sea east across Central Asia to northeastern China

STATUS
Common, but declining in west of range

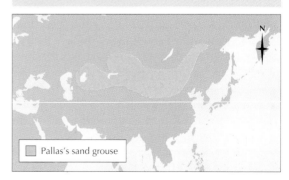

Pallas's sand grouse

sand grouse, *Pterocles burchelli*, although he had ample evidence that birds of this species were nesting in his study area. Sand grouse nest almost year-round in some places. The hen usually lays two to four round eggs, which both parents incubate for 3–4 weeks, males sitting by night and females by day. Between 8 A.M. and 10 A.M. the female lands by the nest. The male gets off reluctantly and the female takes his place. At about 6 P.M. the male returns. The female, by contrast, is eager to be relieved, for she has been sitting in the sun all day and needs to drink. The chicks leave the nest as soon as they dry after hatching, but both parents guard them and bring them food and water.

Water carriers

In 1896 E. G. B. Meade-Waldo published an account of sand grouse carrying water in their feathers to give to their chicks. The idea was viewed with skepticism but is now known to be correct. Male sand grouse walk into water until it reaches their bellies, then crouch down to soak their feathers. They then fly back to the nest with wet belly feathers, and the chicks run forward and take the feathers in their bills. Feathers are usually difficult to wet, but once wetted they hold water quite well. Tests showed that normal feathers hold as much water as a synthetic sponge, but the belly feathers of male sand grouse hold twice as much. The spotted, pintailed (*Pterocles alchata*) and Namaqua (*P. namaqua*) sand grouse, and perhaps other species, carry water like this.

SAND LIZARD

SAND LIZARDS HAVE A wide distribution in Europe and parts of western and central Asia. They are the most common lizard species in many countries such as Hungary and the Czech Republic. They occur in Britain, but there they are restricted to a few areas of heathland in the south or a few areas of sand dunes scattered around the coast.

Sand lizards are rarely more than about 7½ inches (19 cm) in total length, although the largest ever recorded in Britain was just over 8½ inches (22 cm) long. Males are usually slightly bigger than females, and have larger heads. The tail, if it is intact, is slightly longer than the body. Sand lizards vary enormously in color and pattern. Many individuals have a number of rows of oblong dark brown or black markings. These markings usually have an irregular outline, and each almost always has a cream-colored or white spot near its center. Sometimes these dark markings are joined up to form continuous bands.

Other individuals may not have the dark markings, but the small spots remain. In some sand lizards, especially at the eastern edge of their range, these bands may be reddish-brown. Males often have green flanks, and this color becomes more intense during the breeding season. The background color of females is usually some shade of brown.

Not so agile lizard

The name sand lizard is appropriate in England, where it is confined to dry sandy heaths in the south and sand dunes in the northwest. In continental Europe, however, it has a wider range of habitats and is found in woodland clearings, hedgerows and along the borders of fields as well as on dunes and heaths. These habitats are reflected in the names of the sand lizard in French, *lézard des souches* (lizard of tree stumps), and in German, *Zauneidesche* (hedge lizard).

Wherever it is found, the sand lizard prefers dry places such as banks rather than flat ground, and often lives in colonies. Sand lizards can dig well and make their own holes in sandy ground, although they sometimes take over abandoned runs of mice or voles. Sand lizards are timid and dash for the safety of their holes at the slightest disturbance; but they can neither climb

as well nor run as fast as some of their relatives, and so do not deserve their scientific name of *Lacerta agilis* any more than their common name.

Preferred food

Sand lizards feed mainly on insects and spiders but also eat worms, slugs, centipedes and wood lice. Before it is eaten, the prey is killed by means of a vigorous shaking. Larger animals, such as grasshoppers and beetles, have their wings and legs removed first.

Cat fights

In Britain hibernation ends in March and sand lizards start mating in late April, the peak coming in May. During the mating season there is considerable rivalry between males, which defend small territories. Arguments occur if a male enters another's territory. If one lizard is larger than its rival, the smaller quickly retreats, but if evenly matched, the two will fight and blood may flow.

Sometimes the fighting is prefaced by posturing like that of rival tomcats meeting. The sand lizards arch their backs, puff out their necks and lower their heads. They circle about, facing each other, and then attack, rolling over and shaking one another in their jaws. Finally, one sand lizard breaks away and runs off.

A sand lizard in full breeding colors exhibits green flanks.

Sand lizards have short legs and long tails. Continuous bands of dark markings are often also present along the body, as in this male.

SAND LIZARD

CLASS	**Reptilia**
ORDER	**Squamata**
SUBORDER	**Sauria**
FAMILY	**Lacertidae**
GENUS AND SPECIES	***Lacerta agilis***

LENGTH
Usually less than 7½ in. (19 cm)

DISTINCTIVE FEATURES
Rather stocky body; short legs; tail slightly longer than body; color and pattern extremely variable, but male often has green flanks and female generally brownish; may have rows of oblong dark brown or black markings with center spots

DIET
Wide variety of invertebrates, including beetles, spiders, centipedes, earthworms, slugs and wood lice

BREEDING
Age at first breeding: just under 2 years; breeding season: June–July; number of eggs: up to 13; hatching period: about 28 days; breeding interval: 1 year

LIFE SPAN
More than 10 years in captivity

HABITAT
Meadows, field edges, hedgerows, heaths, scrub, sand dunes and open woodland; in Britain, restricted to sandy heaths and coastal sand dunes

DISTRIBUTION
Southern England east through Central Asia, to the east of Lake Baikal, Russia

STATUS
Abundant in most of range; uncommon or rare in parts of western Europe

Courtship is almost as rough as fighting. The females visit the males in their territories and are seized by the males and shaken. Mating then follows. The female lays up to 13 eggs, the size of the clutch depending on the size of the female. The eggs are deposited in a shallow pit dug by the female and afterward covered by leaves or sand. The eggs are laid in June and July and hatch a month later. The young lizards are brown when they first hatch, and before they hibernate in October they have grown from around 2½ inches (6.3 cm) to 3 inches (7.5 cm). They become sexually mature at just under 2 years of age and are fully grown at 4–5 years.

Trusting lizards

Although timid in the wild, sand lizards are easily tamed. The French zoologist Raymond Rollinat has described how he kept lizards on a rockery in his garden. The sand lizards became much tamer than the other species of lizards on the rockery. In 5 days some were tame enough to take food from his fingers and learned to come from their holes when he beat on a tin can. Later the sand lizards would climb over his body and take food from his lips. They did not mind Rollinat moving or even passing them around to his friends. There have been a few people who have claimed to be able to call sand lizards out of their burrows whenever they wished to do so, by making a particular high-pitched whistle. Others who have tried various whistlings have not had any success. Perhaps it has to be a sound that resembles the one made by Rollinat's tin can.

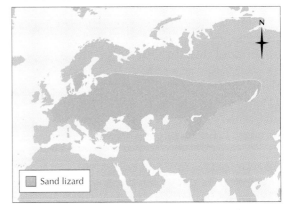

Sand lizard

SANDPIPER

SANDPIPERS ARE SHOREBIRDS OF the family Scolopacidae. Common features of the group include medium-length to long bills and long legs. The hind toe is short, lacking altogether in the sanderling, *Calidris alba*. Winter plumage is usually inconspicuous, but many species have a brighter breeding plumage, often with reddish upperparts and streaked or spotted underparts. The family is divided into several subfamilies, with opinions on the exact number varying between groups of ornithologists. Two of these subfamilies are the Tringinae and the Calidriinae, both of which contain birds called sandpipers: respectively, the tringine and calidritine sandpipers. Thus, sandpiper is a common name given to birds in two groups. Furthermore, it is impossible to define separately from the other members of the Scolopacidae.

Tringine sandpipers include the redshank (*Tringa totanus*) and greenshank (*T. nebularia*) of Eurasia and the greater and lesser yellowlegs (*T. melanoleuca* and *T. flavipes*) of North America. All have checkered, grayish brown plumage, and, as their names suggest, they are distinguished by the color of their legs. The genus *Tringa* also contains the similar green sandpiper, *T. ochropus*. Among the other tringine sandpipers are the terek sandpiper, *Xenus cinereus*, with its up-turned bill, and the common sandpiper, *Actitis hypoleucos*, which breeds in most of Europe, as well as in Central Asia from the Urals to Kamchatka, and overwinters in Africa. Among the calidritine sandpipers are the purple sandpiper (*Calidris maritima*), the knot (*C. canutus*) and the dunlin (*C. alpina*). Some of the smaller species are called stints in Britain or, more widely, peeps. Calidritine sandpipers characteristically have a quiet, twittering call compared with the loud piping of tringine sandpipers.

Long migrations

Most sandpipers breed in the Arctic, often on the coasts and islands of the Arctic Ocean, although the common sandpiper and redshank are two species that breed in temperate climates. Outside the breeding season sandpipers migrate to warmer regions. Some of their migrations cover vast distances. The gray-rumped sandpiper, *Tringa brevipes*, also called the gray-tailed tattler, nests in eastern Siberia and migrates to Australia. The closely related wandering tattler, *T. incana*, nests in Alaska and flies south to Polynesia and occasionally to New Zealand, a distance of 7,000–8,000 miles (11,250–12,900 km). The buff-breasted sandpiper, *Tryngites subruficollis*, has its winter quarters in southern South America and its breeding grounds on the Arctic fringes of northeastern Siberia, Alaska and Canada.

On migration, sandpipers can be seen resting in large numbers on seashores as they travel south and when they return northward. Some habitually feed or roost alone or in twos and threes, but others such as dunlins and knots can be seen in closely packed flocks. When flushed, many sandpiper species fly up together as a flock and wheel about in tight formation before settling together. The usual place to see sandpipers is on the shore or in marshes, where they run nimbly on their long legs. Sandy or muddy shores are preferred, but the purple sandpiper frequents rocky shores. Sandpipers are, however, also found inland, both during the breeding season and during the winter, when sandpipers such as the green sandpiper frequent cress beds, sewage farms and reservoirs.

Surfeit of sea snails

Sandpipers feed on small animals, including insects (which are sometimes caught on the wing), crustaceans, worms, mollusks (such as whelks, mussels and periwinkles) and small fish.

The common sandpiper breeds in most of Europe and parts of Asia. Its North American counterpart is the similar, but slightly smaller, spotted sandpiper, Actitis macularia.

The curlew sandpiper, Calidris ferruginea, nests on the ground in a shallow scrape. The eggs and young are cared for by the female alone, whereas in most sandpipers the male takes a hand.

It has been calculated that a single purple sandpiper can consume 4,600 periwinkles and whelks in a day. Some sandpipers also eat certain plants. In the Arctic, plants may make up most of the diet when insects and other animals are scarce.

Most sandpipers nest near water, either near the seashore or by lakes and rivers, but a few, such as the upland sandpiper, *Bartramia longicauda*, nest in plains and fields, as does the common sandpiper occasionally. Sandpipers do not nest in groups, but redshanks often nest near lapwings, *Vanellus vanellus*. A lapwing, when disturbed, flies up to give the alarm, enabling the redshank to slip quietly off its nest unnoticed. Most species of sandpipers nest on the ground, although the green, wood (*Tringa glareola*), and solitary (*T. solitaria*) sandpipers nest in trees, in the abandoned nests of other birds.

Courtship display

The buff-breasted sandpiper is unique among North American waders in having a lek mating system. Each male defends a lek area into which he attempts to lure females by raising and lowering his wings in display and by performing flutter jumps. Females that venture in are led to slight depressions, known as mating posts, where the male spreads and vibrates his wings and stamps his feet. With wings still spread, the male tilts up his head and emits *tick-tick* sounds. The female enters the male's embrace and rotates so that her back is against his chest. Mating then takes place. Males sometimes try to steal partners by placing themselves between another male and a female. Homosexual mountings also sometimes occur.

BUFF-BREASTED SANDPIPER

CLASS	**Aves**
ORDER	**Charadriiformes**
FAMILY	**Scolopacidae**
GENUS AND SPECIES	***Tryngites subruficollis***

WEIGHT
About 2¼ oz. (63 g)

LENGTH
Head to tail: about 8¼ in. (21 cm); wingspan: 18 in. (46 cm)

DISTINCTIVE FEATURES
Medium-sized shorebird with erect posture; relatively short bill; round head; long, yellowish legs; warm buff face and underparts; brown upperparts with scalelike markings; no wing bar but underwing white

DIET
Flies, beetles and seeds

BREEDING
Age at first breeding: 1 year; breeding season: eggs laid late June; number of eggs: 4; incubation period: 23–25 days; fledging period: 16–20 days; breeding interval: 1 year

LIFE SPAN
Not known

HABITAT
Breeding: high, well-drained tundra. Winter: dry, open ground. Migration: makes stops on short-grass prairies.

DISTRIBUTION
Breeding: Arctic coasts and islands of North America, from Alaska east to Devon Island, Canada; also Wrangel and Ayon Islands, northeastern Siberia. Winter: central Argentina and Paraguay.

STATUS
Scarce; world population: 5,000 to 15,000

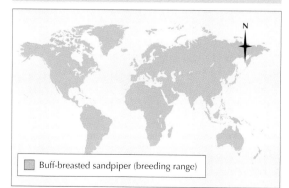

Buff-breasted sandpiper (breeding range)

SAPSUCKER

Several woodpeckers occasionally drill holes in trees and suck the oozing sap, but two North American woodpeckers have made this a habit. These are the yellow-bellied sapsucker, *Sphyrapicus varius*, and Williamson's sapsucker, *S. thyroideus*, which are both similar to other woodpeckers in form. Another type, the red-naped sapsucker, is thought by some to be a closely related species, but others consider it a subspecies of the yellow-bellied sapsucker. The red-naped sapsucker is found west of the yellow-bellied form in the Rocky Mountains.

The yellow-bellied sapsucker is 7½–8½ inches (19–21 cm) long. It has dark upperparts that are mottled with white and pale yellow underparts with a black band on the chest. Its head is black and white with crimson patches on the forehead and throat, although the female lacks the red throat. The red-breasted sapsucker, which is variously considered a subspecies or a separate species, has a completely crimson head and throat. The male Williamson's sapsucker is brightly colored. It is mainly black with two white bars on the head and a white patch on the wing. The chin is red and the belly is yellow. The female is dull in comparison, being mainly brown with black barring and a yellow belly.

In the summer the yellow-bellied sapsucker is common in the woods of southern Canada and the northern United States but is shy and easily overlooked. In the winter it migrates to the southern United States, Mexico and Central America. The red-faced variety is found on the Pacific Coast. Williamson's sapsucker is less common and is confined to the western side of the continent as far north as the Canadian border. It migrates to the southwestern United States and Mexico in winter.

Drilling for sap

Sapsuckers migrate in small parties, flying at considerable heights with the typical undulating, alternate flapping and gliding movement of woodpeckers. Neat, parallel rows of holes in a tree trunk are sure evidence of a sapsucker. A squarish hole, about ½ inch (1.3 cm) across, is drilled through the bark to the phloem (the vascular tissue in plants that transports sugars and other nutrients). The sap flows through the phloem. It contains the products of photosynthesis and is rich in sugars. As the sap oozes out of the hole, the sapsucker sips or laps it with its tongue. When feeding, the sapsucker stands well clear of the trunk and does not hug it as do other woodpeckers. When the flow dries up, the sapsucker just moves around the trunk

A female yellow-bellied sapsucker at the nest hole. Sapsuckers are small North American woodpeckers.

Sapsuckers feed on tree sap as well as the insects that are attracted to the sap. Pictured is a female Williamson's sapsucker.

YELLOW-BELLIED SAPSUCKER

CLASS	**Aves**
ORDER	**Piciformes**
FAMILY	**Picidae**
GENUS AND SPECIES	***Sphyrapicus varius***

WEIGHT
1⅖–2⅕ oz. (40–62 g)

LENGTH
Head to tail: 7½–8½ in. (19–21 cm); wingspan: about 16 in. (40.5 cm)

DISTINCTIVE FEATURES
Small size; dark upperparts with white bars; pale underparts with black breast and fine dark streaking; white rump and wing patches; black-and-white stripes on head and neck; crimson forehead; crimson throat (male only)

DIET
Tree sap, buds, berries, beetles, ants, moths and butterflies

BREEDING
Age at first breeding: 1 year; breeding season: eggs laid late April–June; number of eggs: average 5; incubation period: 12–13 days; fledging period: 25–29 days; breeding interval: 1 year

LIFE SPAN
Not known

HABITAT
Deciduous and mixed coniferous forests; aspen is very important nesting tree

DISTRIBUTION
Summer: from British Columbia in west to Newfoundland, New England and the Appalachians in east; winter: southern U.S., Mexico and Central America

STATUS
Common

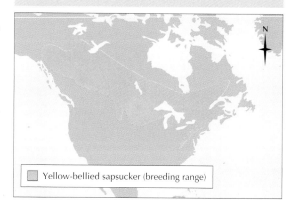

☐ Yellow-bellied sapsucker (breeding range)

and drills another hole. Having exhausted one level, it moves up and starts another row. Sometimes the tree trunk or limb is completely ringed and so dies. Otherwise, the neatly spaced holes appear to do little damage and eventually heal, unless fungus infects them. The amount of damage done by sapsuckers is probably slight, but the holes may leave scars in the wood, so ruining its value as timber. On the other hand, sapsuckers seem to confine their attentions to a small area, feeding regularly at certain trees, so damage appears to be severe. Birches and maples are favored but in the winter the birds feed on a variety of trees and shrubs including poplar, willow, hickory, alder, pines, spruces, firs and fruit trees.

Although well known for feeding on sap, most of the sapsucker's food is insects, which it catches in the air, on trees or on the ground. It also catches insects such as wasps and ants that are attracted to the oozing sap. It beats large insects to a pulp and dips a billful of them into the sap, working them into a ball. In spring sapsuckers eat the buds of trees and occasionally take berries. They usually eat fruit between October and February.

Tapping for an answer

After they arrive in their summer quarters, sapsuckers can be heard tapping at a rate of two or three bursts of taps per minute. They can be attracted by tapping on a tree with a stick. They are also seen flashing about among the trees with tails spread, calling at the same time. Males arrive at the breeding grounds before the females, which tend to winter farther south. After they have paired the male and female take turns excavating a nesthole in the trunk of a dead tree, often an aspen. The entrance is about 2 inches (5 cm) across and the hole may be over 1 foot (30 cm) deep and 5 inches (12.5 cm) in diameter at the widest. The eggs are pure white, as is usual in woodpeckers, and the average clutch is 5, although the clutch size increases from the south to the north. Both sexes share incubation, which takes 12–13 days. After they have left the nest, the fledglings fly to one of the feeding trees and the parents continue to feed them for up to 2 weeks.

Food for all

Sapsuckers may be a nuisance to fruit farmers or foresters, but their tree-drilling habits are a benefit to many other animals, which come to feed on the sap. Frequent visitors include hummingbirds, other woodpeckers, nuthatches and warblers. Chickadees, flying squirrels and porcupines can also be seen at sapsucker holes. Many insects are also drawn to the sap, such as flies, moths, butterflies and beetles.

Evenly spaced rows of holes in a tree trunk or large branch are a sure sign that a sapsucker has been feeding in the area. An immature yellow-bellied sapsucker is pictured.

SAWFISH

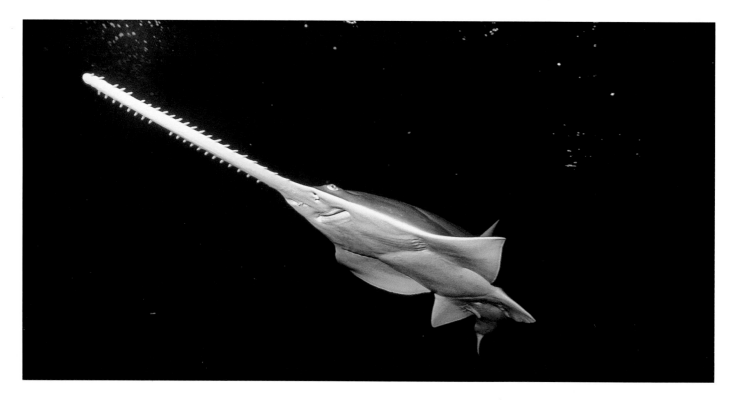

The sawlike snout that gives the sawfish its common name may represent up to one-third of its total body length. Despite its sharklike shape, the sawfish is more closely related to rays.

THE SAWFISH, OF WHICH there are seven species, look like sharks but behave like rays and are related to the guitarfish. Sawfish have a sharklike tail, a somewhat flattened body and pectoral fins that are not joined to the sides of the head as they are in rays and guitarfish. However, in common with these species the gill openings are on the underside of the head. The snout is drawn out into a long, flattened blade, the "saw" from which the fish's common name is derived, with a row of strong teeth sticking out sideways on each side. These projections are not the true teeth, although they look like teeth and have dentine, enamel and a pulp cavity. They more closely resemble the dermal denticles, or skin-teeth, that are characteristic of sharks and rays. The flat blade of the saw, known as the rostrum, is made of cartilage, like the rest of the skeleton, and the teeth are contained in deep sockets within this.

Sawfish may grow up to 25 feet (7.5 m) long and occasionally a specimen measuring 30 feet (9 m) or more is recorded. They may weigh more than 5,000 pounds (2,268 kg) when fully grown.

Sawfish live in coastal waters in all warm seas down to depths of about 30 feet (9 m) and sometimes even penetrate well up rivers into fresh water. They enter estuarine areas prior to giving birth. The smalltooth sawfish, *Pristis pectinatus*, of the Gulf of Mexico, is found far up the Mississippi River, and there is also a population in Lake Nicaragua. Another species, the large-tooth sawfish, *P. perotteti*, goes well up the Zambesi River, and a third species, the knife-tooth sawfish, *Anoxypristis cuspidata*, swims up the larger rivers of the Indian subcontinent.

The family of saw sharks, Pristiophoridae, contains five species. These are true sharks with gill openings on each side of the head. They are not more than 4 feet (1.2 m) long and the teeth in their saws are alternately large and small.

Killing with the saw

It is likely that sawfish occur in large numbers on muddy river bottoms. In the sea, where food is more readily abundant, they are probably even more numerous. Sawfish use their saws to disturb the mud of the river bottom in order to search for mollusks, crustaceans and sea urchins. Their true teeth are small with blunt crowns and are set in rows in the jaws, forming a mill. These teeth are used for crushing and grinding prey such as shellfish.

Sawfish employ an alternative method of feeding, similar to that used by sailfish, in which they swim into a shoal of smaller fish and strike to the left and the right with their saws, killing and stunning the surrounding prey, which they are then able to consume at leisure. Scientists have discovered that the sight of a fish nearby automatically causes the sawfish to move its saw from side to side.

SAWFISH

CLASS	**Chondrichthyes**
SUBCLASS	**Elasmobranchii**
ORDER	**Pristiformes**
FAMILY	**Pristidae**
GENUS	***Anoxypristis*** and ***Pristis***
SPECIES	**7, including smalltooth sawfish,** ***Pristis pectinata*** **(described below)**

WEIGHT
Up to 770 lb. (350 kg)

LENGTH
Up to 25 ft. (7.5 m)

DISTINCTIVE FEATURES
Dark gray to blackish-brown upperparts; white to yellowish underparts; pectoral fins tipped with black on ventral surfaces

DIET
Mostly fish; crustaceans, other invertebrates

BREEDING
Viviparous. Number of young: up to 23; gestation period: up to 12 months.

LIFE SPAN
Several years

HABITAT
Inshore coastal waters and islands; lakes, bays, lagoons; estuaries and rivers

DISTRIBUTION
Western Atlantic: North Carolina, Bermuda and northern Gulf of Mexico south to Brazil. Eastern Atlantic: Gibraltar to Namibia; possibly Mediterranean Sea. Indian Ocean and western Pacific: Red Sea south to South Africa and Myanmar (Burma); possibly to Thailand, Philippines and Northern Territory of Australia. Possibly also eastern Pacific.

STATUS
Endangered

Smalltooth sawfish

Soft-saw young

The females give birth to free-swimming young, the eggs being incubated within the body and hatching just before they leave it. The young are born with soft saws through which the teeth barely project. Moreover, the teeth are covered with a membrane until the baby sawfish is born.

The number of young born at one time can be gauged from the few instances in which pregnant females have been caught and examined. One, estimated to weigh 5,300 pounds (2,400 kg), carried several young. Another sawfish, 15½ feet (4.7 m) long, was found to contain 23 young when it was examined.

More prey than predator

In some parts of India sawfish are regarded with great trepidation, because it is believed they will attack a human. While this is possible, it is true to say that any large animal may kill a human accidentally; moreover, sawfish are far more likely to be killed by humans than vice versa.

In the late 18th century, one fisher reportedly took 300 sawfish from a Florida river system in one season. However, today the global sawfish population is dwindling. Trade in sawfish fins and skin, both of which are used in medicines and soaps, has greatly reduced their numbers. The saws are sold as curios in tourist shops, and those collected in Malaysia are sold in China for use in traditional medicines. Commercial and recreational fishing, water pollution, and reduction in the populations of prey species have all combined to adversely affect sawfish populations. Like sharks, sawfish are highly vulnerable to overexploitation because they mature slowly and produce relatively few young. Today all sawfish species are regarded as endangered.

The smalltooth sawfish is the most numerous sawfish in the United States, but its numbers have fallen in recent times. It may no longer occur on the East Coast, where it was once abundant.

SAWFLY

Sawflies often have a beelike or wasplike coloration, and some species are called wood wasps. The markings of this harmless wood wasp mimic those of a true, stinging wasp as a means of self-defense.

THE SAWFLIES ARE NAMED after the sawlike tip of the ovipositor with which the female drills holes in plants to lay her eggs. Sawflies are related to ants, bees and wasps but can be distinguished from them by not having a narrow wasp waist. However, some have striped bodies like bees and wasps, and the long ovipositor may be mistaken for a stinger. The black-and-yellow horntail, *Sirex gigas*, for example, is ¾ inch (2 cm) long and is frequently mistaken for a wasp. Some sawflies, including the horntail, are called wood wasps because they lay their eggs in wood rather than on leaves or on the other soft parts of plants.

The larvae of sawflies are easily mistaken for caterpillars. However, close examination shows that as well as the three pairs of true legs, they have at least six pairs of false legs, or prolegs. Caterpillars of moths and butterflies, on the other hand, never have more than five pairs of prolegs. In addition, sawfly larvae have only one pair of minute eyes, whereas caterpillars have a cluster of four to six eyes on each side of the head.

Drilling for a living

Adult sawflies are very short-lived, the males dying shortly after mating and the females surviving only long enough to lay their eggs. The eggs are laid in slits made in plants by the ovipositor. The ovipositor is made up of two thin, rigid blades, protected when not in use by a pair of sheaths. The lower parts of the blades are notched to form saws or rasps. The two blades work alternately, moving to and fro as they drive into the plant tissue. The larvae of sawflies feed on many plants, including pines, turnips, roses and apples, and are often serious pests.

The larvae are found in a number of places, such as in leaf galls, inside wood, in the open on leaves or living socially in silken webs. Those that live on leaves either scrape away the cuticle covering the leaves, forming conspicuous patterns as may be seen on roses, or, like the pear sawfly, reduce the leaves to skeletons. The larva of the pear sawfly, *Eriocampa limacina*, is often called the slugworm and may be mistaken for a slug. The similarity is increased by the shiny slime it secretes, apparently as a deterrent to insect-eating birds. The pear sawfly larva assumes the conventional form of a sawfly larva for only a short time before it pupates.

The larva of the palisade sawfly, *Nematus compressicornis*, which lives on poplars, erects a barricade around itself. It lowers its mouth to the leaf and exudes a drop of saliva. As the head is raised, the drop is drawn out and immediately sets to form a fine, shiny post. The action is repeated until the larva is surrounded by a fence. Having eaten the leaf within its enclosure, the fence is eaten and the larva moves to another leaf and erects another palisade. The fence is very flimsy and it is difficult to imagine what protection it provides for the larva.

Burrows and cocoons

Sawfly larvae may take several years to mature. Some drop to the ground and burrow to pupate, whereas others spin a cocoon attached to a leaf or twigs. Larvae that feed in the wood pupate in the tunnels they have made. The jumping-disc sawfly, *Phyllotoma aceris*, pupates in a cocoon that also serves as a parachute. The larva lives in sycamore leaves and feeds on the tissue of the leaf without damaging the upper or lower cuticle. When full grown, it cuts a series of perforations in the upper cuticle to form a circle. Then it weaves a silk hammock from the inside of this circle. The hammock does not touch the lower cuticle. Once safely inside, the disc of upper cuticle separates and drops to the ground with the larva attached by its hammock. When it has landed, the larva works its way into a sheltered space by jerking its disc along, a fraction of an inch at a time.

Many sawfly larvae are strikingly colored. Some are black and white, while others are black and yellow. This is apparently a warning coloration, as some sawflies can squirt irritating fluids from glands on their undersides. The pupae are hunted by birds and rodents. In the United States scientists found that the proportion of pupal cocoons that were damaged was a good indication of the size of the local mouse population. In Britain 100 pupae were once found in the stomach of a red squirrel. Both larvae and pupae

are parasitized by ichneumon wasps, such as those belonging to the genus *Megarhyssa*, which have ovipositors that can bore through 3 inches (7.5 cm) of wood.

Controlling sawfly

Many sawfly larvae are serious pests and have spread beyond their native homes. The wheat-stem borer, *Cephus pygmaeus*, for example, has been carried to North America, and several wood wasps have been spread, perhaps in imported timber. In the early 1940s the European spruce sawfly, *Gilpinia hercyniae*, reached eastern Canada and became a serious threat to forests there. A search was quickly made for predators, such as ichneumons, which could be imported to combat it. Suitable parasites were found and liberated in the forests. However, the sawflies were suddenly killed off by a disease and the parasites were not needed.

Another wood wasp reached Australia in about 1947. Apart from holes made by the larvae, the wood wasp caused great damage because the females injected a fungus into the plants with their eggs. Experiments into ways of killing wood wasps are currently underway. These include the introduction of ichneumons and certain nematode worms. The worms sterilize the female wood wasps so they lay eggs packed with young nematodes, which also eat the destructive plant fungus.

SAWFLIES

PHYLUM	**Arthropoda**
CLASS	**Insecta**
ORDER	**Hymenoptera**
SUBORDER	**Symphyta**
FAMILY	**Tenthredinidae; Siricidae (wood wasps and horntails); Cimbicidae; others**

ALTERNATIVE NAMES
Wood wasp, horntail (Siricidae only)

LENGTH
Tenthredinidae: 2.5–15 mm; Siricidae: more than 14 mm; Cimbicidae: 4–30 mm

DISTINCTIVE FEATURES
Adult: small size; often resembles wasp but without waist and highly social behavior; long, sawlike ovipositor (female only). Larva: resembles butterfly or moth caterpillar.

DIET
Adult: pollen and nectar. Larva: plant matter and wood.

BREEDING
Eggs laid in plant tissue or wood in holes made by female; larvae hatch and undergo full metamorphosis

LIFE SPAN
Adult: few days. Larva: up to several years.

HABITAT
Inside plants or on leaves

DISTRIBUTION
Worldwide except polar regions

STATUS
Common

Sawfly larvae are voracious eaters. In their characteristic feeding position, posterior end hanging down, they strip the leaves of a plant such as rowan (below).

SCAD

SCAD IS THE ORIGINAL NAME for the horse mackerel. Although it bears a close resemblance to a mackerel, the scad differs in having its lateral line covered from head to tail with a row of flat, bony plates, those along the hind part of the body being spiny. The head is large, the first dorsal fin is spiny, the second is long and the tail fin is slightly forked. The anal fin is long, and its first two spines are separated from the rest of the fin. The pelvic fins lie forward, below the medium-sized pectorals. The body is compressed from side to side, up to 16 inches (40 cm) long, bluish gray in color with a greenish tinge on the back, silvery on the flanks and white on the belly.

Some of the other 140 species in the family Carangidae are known as jacks. These include the amber jacks, which lack the bony plates over the lateral line, and the greater amberjack, *Seriola dumerili*, of the Atlantic. The yellowtail amberjack, *S. lalandi*, is a Pacific relative that lives off the coasts of Mexico and California. Related to all these are the pompanos, such as the Florida pompano, *Trachinotus carolinus*, of the North American seas and the lookdown, *Selene vomer*, of the North Atlantic, which has eyes set so high up on its head that it seems to be permanently looking down. The pilot fish, *Naucrates ductor*, is also related to scad.

The scad and two other species of the same genus, *Trachurus*, live mainly in European waters, including the Mediterranean and Black Seas.

Carnivores from hatching

The scad, which moves about in large shoals, lives in coastal waters in summer and retires to deeper water, down to 330 feet (99 m), in winter. Young scad in particular are often numerous in summer in shallow seas with sandy bottoms. They feed on other fish, especially herring, sprat, pilchard and anchovies, as well as squid and crustaceans. In the North Sea and in similar areas they spawn from June to August, with a peak in July. The eggs, 0.8 millimeters in diameter, contain a reddish oil droplet, enabling them to float near the surface. On hatching, the larvae are 2.5 millimeters long. They do not grow their fins until they are four times this length. They feed at first on diatoms, copepods and crustacean larvae, fish eggs and the larvae of other species. The following winter their diet changes to mainly small fish and euphausians (small crustaceans).

Many members of the family Carangidae, including jacks, cobias and remoras, are important food fish and support large commercial and recreational fisheries throughout their range.

SCAD

CLASS	**Osteichthyes**
ORDER	**Perciformes**
FAMILY	**Carangidae**

GENUS AND SPECIES **Atlantic horse mackerel,**
Trachurus trachurus

WEIGHT
Up to 4⅖ lb. (2 kg)

LENGTH
Up to 28 in. (70 cm)

DISTINCTIVE FEATURES
Slender, elongated body; heavy, long head; dark gray-blue back with greenish tints; silvery sides with golden flush; white ventral area; dusky spot on edge of gill cover; series of wide, bony scales along lateral line, flexible in front, sharp edged and hard toward tail; first dorsal fin has long spines united by membrane, separate from second dorsal fin

DIET
Fish, crustaceans, cephalopods

BREEDING
Age at first breeding: 3–4 years; breeding season: July–August (in Celtic Sea). Young fish, ⅖–2¾ in. (10–70 mm) long, found beneath large pelagic jellyfish.

LIFE SPAN
Several years

HABITAT
Coastal areas with sandy substrate

DISTRIBUTION
Eastern Atlantic: Iceland south to Senegal, including Mediterranean, Marmara and Black Seas. Also western Atlantic Ocean.

STATUS
Common

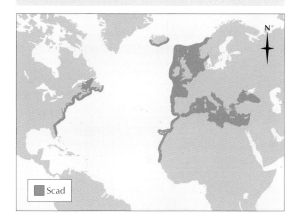

Scad

Fish meal fisheries

There are large commercial fisheries for scad off Spain and Portugal, but elsewhere they are caught only incidentally with other catches, since they are not considered particularly palatable and are used only for fish meal, especially the young scad. Their growth rate has been studied in Spanish waters: the young fish reach 3½–8 inches (8.75–20 cm) at the end of their first year, 8–10 inches (20–25 cm) at 2 years, and a length of 10½ inches (26.25 cm) in their fourth year.

Jellyfish sanctuary

Little is known of the predators of scad, but scientists presume that when they reach adult size they are eaten by predatory fish, dolphins and porpoises. When they are 1¼–1¾ inches (3.1–4.4 cm) long, they regularly seek refuge under the bells of jellyfish. This habit has already been described with regard to the small fish *Nomeus albula* in the entry for Portuguese man-of-war, and it is of interest that the scad should have the same habit, although it associates with completely different kinds of jellyfish. The young scad shelter under the medusoid, or bell-shaped jellyfish, which include the common jellyfish, *Aurelia aurita*, the blue jellyfish, *Cyanea capillata*, and the barrel jellyfish, *Rhizostoma octopus*. The scad swim just in front of the jellyfish they have adopted and, when alarmed, dive for protection among its tentacles, although scientists believe that they eat their protector's tentacles, and perhaps even its eggs. It is possible, although this has not been proved, that young scad eat other small animals that also shelter under the jellyfish.

Also called the akule, the bigeye scad, Selar crumenophthalmus, gathers in large shoals, especially during the spawning season, making it an attractive catch for fishers.

SCALARE

The scalares' common name of angelfish derives from their winglike dorsal and ventral fins. The three scalare species should not be confused with the marine angelfish.

THE SCALARE, OR FRESHWATER angelfish, is a popular aquarium fish, known for its ability to keep stock still. Its body is markedly flattened from side to side, and it is only a little longer than it is deep, even without the very tall dorsal fin and the equally prominent anal fin. Two types of marine fish are also known as angelfish: the monkfish, genus *Squatina*, and the many members of the family Pomacanthidae.

The body is gray green with a silvery sheen and pale reddish brown blotches. There are often darker markings on top of the head. Black bands run across the flanks. One of these extends up into the dorsal fin and down into the anal fin. The delicate tail with its two points and the long filamentous point on the anal fin give an impression of movement. The pelvic fins also are long, thin and backward-flowing. This gives the scalare, even when stationary, the appearance of being in a state of arrested motion. The most common aquarium scalare is about 2 inches (5 cm) long, but the fish may be up to 6 inches (15 cm) long and 10 inches (25 cm) high, including fins.

The three closely related species of scalares, which may actually be subspecies of the same species, live in the Amazon, Orinoco, Negro and Tapajoz Rivers of northeastern South America. The three forms are usually known as the angelfish, *Pterophyllum leopoldi*; the lesser or freshwater angelfish, *P. scalare*; and the deep angelfish, *P. altum*. They differ in shape and color but readily interbreed, producing fertile offspring.

Camouflage stripes

The body of a scalare is so flat that when the fish is viewed head-on it seems to have height but no breadth. When viewed broadside-on, the dark bands serve the same purpose as the stripes of a tiger, making the fish inconspicuous among water plants. This is probably why the scalare, although it tends to swim little for much of its time, giving the impression of habitual immobility, is easily frightened by a sudden movement when it is in a tank with no water plants. In such circumstances it is likely to dash itself against the glass sides in panic.

Scalares eat small animals, such as water fleas, freshwater shrimps, various insect larvae, worms and sometimes tiny fish.

Devoted parents

The difficulty of telling a male scalare from a female is increased by the fact that in the breeding season both develop a genital papilla. The males at this time may make creaking sounds with their jaws, either at a rival male or as part of the courtship. The female lays her pale yellow, oval eggs, about 1 millimeter long, on the broad leaf of a water plant, which she and her mate first clean of small algae and grains of sediment. They both fan the eggs continually during the 24–36 hours before hatching. The young fish are helped out by their parents, which chew at the egg casings, and as each youngster leaves, one of the parents takes it in its mouth and spits it onto another leaf, where it hangs by a short thread. The wriggling of the infant fish as they hang from the leaf helps to develop their muscles and circulates the water around them.

The parents continue to care for their brood for 4–5 days, during which time they remain suspended from the leaf. If one of the young fish falls from its thread, one of the parents swims down, picks it up in its mouth and spits it back onto the leaf.

At first the newborn scalare is tadpolelike, slightly more than 3 millimeters long, with a well-filled yolk sac on which it feeds during its

SCALARES

CLASS	**Osteichthyes**
ORDER	**Perciformes**
FAMILY	**Cichlidae**

GENUS AND SPECIES **Lesser angelfish, *Pterophyllum scalare* (described below); deep angelfish, *P. altum*; angelfish, *P. leopoldi***

ALTERNATIVE NAME
***P. scalare*: freshwater angelfish**

LENGTH
Up to about 6 in. (15 cm)

DISTINCTIVE FEATURES
Strongly compressed body; long, flowing dorsal, anal and pelvic fins; green-gray sides with brilliant silver sheen; paler belly; four bold transverse bands of deep black or gray

DIET
Small crustaceans, worms and insect larvae; sometimes tiny fish

BREEDING
Number of eggs: up to 1,000; hatching period: 24–36 hours

LIFE SPAN
Not known

HABITAT
***P. scalare*: quiet, weedy rivers**

DISTRIBUTION
***P. scalare*: Guiana south to Amazon Basin.
P. altum: Orinoco and Negro Rivers.
P. leopoldi: Guyana and Negro River.**

STATUS
Generally common

Scalares

first days. By the time it is 12 days old and ⅓ inch (8.5 mm) long, the juvenile scalare resembles a miniature minnow and is quite unlike the adult. At 20 days old and ½ inch (1.3 cm) long, the body begins to deepen. Eight days later, the fins are noticeably larger and the pelvic fins are growing

long. By the time it is 36 days old and ¾ inch (2 cm) long, the young scalare has taken on almost the same shape as its parents.

Few predators?

Although they are such popular aquarium fish, scalares have not been widely studied in their native habitat. However, it is likely that they have few predators. They live in quiet streams and backwaters with plenty of water plants, where the fish stay almost motionless for hours on end. Their black stripes look very similar to shadows among the vertical stems, and, enhancing the camouflage effect, the fish's fins sway gently, like water plants in a slight current.

Confusion over naming

When aquarium keepers first started to collect scalares, the fish were known as angelfish. However, there were other fish already bearing this name, notably the marine angelfish, which are equally popular aquarium fish. To avoid confusion the name scalare was adopted, after the scientific name of one freshwater angelfish species, *Pterophyllum scalare*.

Because these fish breed so readily several aquarium varieties have evolved, one being melanistic (darkly pigmented). These are known as black angels.

The scalares' dark transverse bands and ability to remain near-motionless for some time provide the fish with excellent camouflage among water plants.

SCALE INSECT

Scale insects, such as Parthenolecanium corni (above), are serious plant pests. One female can produce 30 million offspring in a year.

SOME SCALE INSECTS ARE counted among the most serious of agricultural pests, yet no family of insects has provided more raw materials for human use. Along with mealy bugs they make up a family of insects related to aphids and cicadas. All are flattened and degenerate (lacking certain ancestral characteristics) and are covered with a secretion of wax or resin, which in most forms is mealy or cottony. Some are covered with a scale and look like tiny limpets on the twigs of trees and bushes. Most scale insects are small, usually about ⅕ inch (5 mm) long. They range worldwide, apart from the polar regions, having been carried about on plants or on parts of plants by humans.

Degenerate insects

The greatest damage to crops is done by female scale insects, which infest vegetation in very large numbers. All females are wingless, and many are quite degenerate as adults, lacking legs and with reduced antennae. The males are more normal, although some are wingless and others have only one pair of wings, the second pair being very small and slender with one or two hooks to fasten them to the front wings. All males lack mouthparts and therefore do not feed, and are never present in such numbers as the females. In some species there are two types of males, winged and wingless, which alternate from one generation to the next. In most species,

however, males are rare and the females reproduce by parthenogenesis (development of an egg without fertilization by a male).

Plant pests

Scale insects have a proboscis, or sucking tube, which is pushed into the tissues of a leaf or root of a plant to suck up sap. This harms the plant, and when scale insects occur in large numbers the results can be devastating. One example is the cottony cushion scale, *Icerya purchasi*, introduced into California from Australia and New Zealand in 1868. It wiped out hundreds of thousands of orange trees in the following 25 years. Its numbers were controlled only by introducing its natural predator, a ladybug, from Australia. The elm scale, *Eriococcus spurius*, of Europe and North America is another well-known pest. It kills trees by sucking the sap from the undersides of branches. Some species, such as the pine leaf scale, *Chionaspis pinifoliae*, of North America, attack only certain types of plants. The heath scale, *E. devoniensis*, of Britain forms galls on heaths. Others, such as the mussel scale, *Lepidosaphes ulmi*, feed on 130 different species of trees and bushes.

Well-protected eggs

Some scale insects lay eggs, others give birth to active young, as in aphids. In both cases, the larvae are flattened and oval in outline and

SCALE INSECTS

PHYLUM	**Arthropoda**
CLASS	**Insecta**
ORDER	**Hemiptera**
FAMILY	**Coccidae**

GENUS AND SPECIES **Pine leaf scale,** *Chionaspis pinifoliae*; **cochineal insect,** *Dactylopius coccus*; **Chinese wax insect,** *Ericerus pe-la*; **heath scale,** *Eriococcus devoniensis*; **elm scale,** *E. spurius*; **cottony cushion scale,** *Icerya purchasi*; **mussel scale,** *Lepidosaphes ulmi*; **others**

LENGTH
Usually about ⅕ in. (5 mm)

DISTINCTIVE FEATURES
Female wingless, often legless; antennae reduced; hard waxy scale covering; waxy or white secretions; male lacks mouthparts

DIET
Plant sap

BREEDING
Mostly reproduce by parthenogenesis or live young produced without mating; female may produce 30 million eggs a year

LIFE SPAN
About 4 weeks

HABITAT
Plants

DISTRIBUTION
Worldwide except poles

STATUS
Abundant; sometimes superabundant

therefore are not easily seen on stems or leaves. At first they have legs and crawl about actively, but in many species the legs are soon lost. In all species the larvae soon settle in one spot, covering themselves with the secretions characteristic of their species. Scale insect eggs are not laid in the open but underneath the female's body, under the scale, or covered with the cottony or mealy secretions. In some species, the eggs do not hatch until the female is dead, her body protecting them during the winter. In one subfamily, the Ortheziinae, the females have a covering of small, white, waxy plates and a wax scale over the eggs at the end of their abdomens.

Although ladybugs and other predatory insects are the main predators of scale insects and mealy bugs, there are many others that are, in a sense, accidental predators. These are the many insects that feed on the surface tissues of plants, chewing them with their jaws. They are known to take animal food readily, even though they are basically vegetarian, and chew any scale insects that happen to lie in their path.

Universal providers

The earliest recorded use made of the secretions of scale insects is of the quantities of sweet honeydew given out by the genus *Trabutina* in Palestine. The honeydew solidifies on the leaves and on the ground beneath and forms the manna on which the Children of Israel are said in the Bible to have fed for 40 years.

Also well-known and much more widely used since the 16th century is the resinous secretion given out by *Laccifera lacca*. This scale insect feeds on banana, fig and other trees from India to the Philippines and Taiwan. Its secretion has been used in shellacs, varnishes, polishes and sealing wax, as well as for making inks, buttons, phonograph records, pottery, linoleum, airplane coverings and electrical insulation, to name just some of the main products. It takes 150,000 insects to produce a pound of lac, and, until synthetic products began to take its place, anything up to 90 million pounds (41 million kg) of lac were collected each year.

Another well-known product from scale insects is cochineal, at one time used in dyes, medicines, cosmetics and confectionery. In addition, the Native Americans used to make a kind of chewing gum from the wax given out by a scale insect, and the Chinese used the wax from another species for making candles.

The cochineal insect, genus Dactylopius, *produces the natural red dye cochineal, some of which can be seen on the white card in the photograph above. The bugs are also used to control the spread of the prickly pear.*

SCALLOP

Scallops have many brightly colored eyes lining the gape of their shells. Pictured is the queen scallop, Chlamys spercularis, *which ranges from Europe to northern and eastern Africa, as far south as Sierra Leone.*

EXCEPT FOR THE file-shells, family Limidae, scallops are the only bivalves that swim. They also have many highly developed eyes. The great scallop or St. James' shell, *Pecten maximus*, is almost circular in outline, with "ears" on the shell near the hinge. The right or lower valve is convex and slightly overlaps the left or upper valve, which is flat. The generic name *Pecten* is from the Latin, meaning a comb, and refers to the ridges or corrugations that run from the hinge to the margins of the shell and which look like the teeth of a comb. The edges of the valves are wavy, and this pattern has given the name "scalloped" for a particular method of cutting the edges of fabrics. Although most scallops have one flat valve, some, including *Scaeochlamys lividus* of Australia and the members of the genera *Chlamys* and *Aequipecten*, have both shells convex. Scallop shells tend to be colored a delicate pink, red or yellow, and the corrugations of some species have frills, overlapping plates or spines.

The 300 or more species of scallops occur throughout the world, from shallow water just below the lowest tidemark, to about 330 feet (100 m). They are sometimes found between tidemarks when stranded, and there is at least one species that lives between tidemarks.

Swimming, walking and sitting shells

A scallop normally lies free on the seabed and will swim away when a predator comes near. They seem to swim at other times for no particular reason, and scallops have occasionally been caught swimming just under the sea's surface.

Some species can fasten themselves to a solid support by a byssus or beard, a bunch of horny threads (see mussel, discussed elsewhere). The small or southern scallop, *Pecten latiauritus*, of the Pacific Coast of North America, fastens on the fronds of kelp, but it can also travel by pushing out its foot, pressing the tip against a solid surface, and then contracting it violently, thereby hitching itself along.

Some scallops cement themselves to rocks or other shells, for example the hoofed shells *Chama* of North America. The purple-hinged scallop, *Hinnites giganteus,* of North America is free while young, but cements itself to a rock when it has grown to 1 inch (2.5 cm) long.

Swimming backward or forward

Normally a scallop lies on the seabed with its shell slightly agape and the space between the valves curtained by the edges of the mantle and its tentacles. All around the gape, nestling among the tentacles, are 100 or more blue eyes. In the way they feed and reproduce, scallops are like other bivalve mollusks such as mussels and oysters. They differ in being able to move about, swimming backward or forward.

When a scallop has opened its shell and taken in water, its mantle edges come together except at one point at the front. If the valves are brought together quickly, water is forced out through the opening left at the front and the scallop is driven backward. Alternatively, the mantle edges may close, leaving two openings at the rear, one on either side of the hinge. When water is driven through these openings, the scallop is driven forward.

In other bivalves the ligament of the hinge is on the outside and must contract to pull the shell open when the adductor muscles are relaxed. In scallops the ligament is on the inside and expands to push the shell open. Scallops also differ from some other bivalves in that instead of having two muscles, one large and one small, running between the valves, there is only one very large, powerful muscle. This is the part of the scallop body used in cooking.

Limited maneuvers

The scallop is a relatively powerful swimmer, although it is somewhat erratic in its side-to-side movement. Using first one opening near the hinge and then the other, the scallop can turn. By varying the use of the temporary openings formed by the edges of the mantle, the scallop can make other limited maneuvers in the water.

SCALLOPS

PHYLUM	**Mollusca**
CLASS	**Bivalvia**
ORDER	**Ostreoida**
FAMILY	**Pectinidae**

GENUS AND SPECIES   **Over 300 species, including the great scallop, *Pecten maximus***

ALTERNATIVE NAMES
St. James' shell; Coquille St. Jacques

LENGTH
Up to 6 in. (15 cm)

DISTINCTIVE FEATURES
Bivalve; one flat and one rounded, strongly ribbed shell; numerous eyes

DIET
Filter feeder, mainly on phytoplankton

BREEDING
Hermaphroditic, releasing sperms and eggs into water; planktonic larvae hatch; settle as tiny bivalves on hydroids, bryozoans and seaweeds; release themselves after a few months and settle on seabed

LIFE SPAN
Up to 20 years

HABITAT
Coarse sand and gravel in shallow water

DISTRIBUTION
Worldwide

STATUS
Common

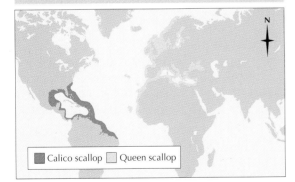

Calico scallop   Queen scallop

One way to observe the power that this mollusk can muster when swimming is to hold a scallop out of water, with the finger and thumb over the hinge region. The scallop opens its shell slightly then shuts it violently, making a sound like a cough. In water this same movement creates a strong jet of water.

Sense organs

Each of the scallop's eyes has a lens, iris, cornea and double retina. It begins as a group of pigmented cells that form into a cup and ends in a structure comparable with the complex eyes of vertebrates. The growth of new eyes is common, so the total number is always changing. If a scallop loses its eyes through injury, they can be replaced by new ones. A scallop that loses all its eyes can regrow them in two months.

The tentacles around the edges of the scallop's mantle are sensitive organs of touch. A scallop also has a well-developed chemical sense that is both taste and smell. In addition a scallop has an organ of balance: a sphere of cells with sensitive protoplasmic hairs pointing into the cavity of the sphere in which lies a small, round mass of calcareous matter. As the scallop moves, the stone changes its position, touching the hairs and enabling the scallop to establish whether it is leaning to the side or falling upside down. If necessary, a scallop will leap to land the right way up, on its convex valve. A scallop is said to be able to jump to right itself even when it is out of water.

Alert for predators

The predators of scallops are mainly starfish and octopuses. The approach of either of these animals sends a scallop swimming away. Even the juices from the body of a starfish poured into an aquarium containing scallops will set them on the move. They perceive the presence of predators by their chemical sense and by a shadow falling across their eyes. However, there is still scientific debate as to why these mollusks should have so many sense organs compared to other mollusks, and particularly why they should have so many highly organized eyes.

As with all scallops, the shell of this doughboy scallop, Chlamys asperrimus, *of New South Wales, Australia, is opened and closed by a single large muscle, the adductor.*

SCALY-TAILED SQUIRREL

The flightless scaly-tailed squirrel, or Cameroon scaly-tailed squirrel, is the only scaily-tailed squirrel without a flight membrane, and superficially resembles a large dormouse.

SCALY-TAILED SQUIRRELS ARE also called scaly-tailed flying squirrels, but are not true flying squirrels, and are not even true squirrels. In fact, these gliding animals are quite primitive rodents, in some ways resembling the hypothetical ancestors of rats, squirrels and porcupines. Except for one species, the flightless scaly-tailed squirrel, all have a flight membrane of skin running from the forelegs to the hind legs and from there to the base of the tail. In front it is strengthened by a rod of cartilage running back from each elbow, whereas in true flying squirrels the flight membrane is supported by a bone protruding from the wrist. The long, fine fur conceals a slender body. The color of the fur is variable, even within species, but is usually gray or brown above and white or yellowish on the underparts. Pel's scaly-tailed squirrel, *Anomalurus peli*, has bold, black-and-white coloration, and Beecroft's scaly-tailed squirrel has a warm orange tint to the underside. At the base of the tail, on the underside, is a double row of overlapping, keeled scales with sharp, backwardly directed points. These act as climbing irons, supporting the animal in its characteristic vertical pose on tree trunks, allowing it to hang passively from the long claws on its hands. Scaly-tailed squirrels also have large ears, usually large eyes and long whiskers. There are seven species, all of which live in the rain forests of West Africa and central Africa.

Devoted to trees

Scaly-tailed squirrels live in tall trees, spending most of their lives up in the canopy, 100 feet (30 m) or more above the ground. They are so specialized for the arboreal life that if they find themselves on the ground, they move awkwardly, hampered by their flight membranes. They rest by day in their dens in hollow trees, sometimes coming out in the late afternoon to bask in the sun or in the evening to feed. They can run nimbly along branches but can also travel from one tree to another by flying leaps. They launch themselves into the air, spread all four legs to stretch the webs and glide, typically for 60 feet (18 m), and occasionally much farther. To land they lift the front part of the body until it is almost vertical. They climb vertical trunks with a series of short leaps, first the forelegs together and then the hind legs. Meanwhile, the tail with its double row of scales is pressed against the trunk for support.

Tree surgeons

The scaly-tailed squirrels are bark-eating specialists, gnawing bark from a dozen or so preferred species of food tree, including miombo (*Brachystegia* spp.) and ironwoods (*Cynometra* spp.). These large rain forest trees are members of the pea family. The scaly-tailed squirrels supplement their diet with the nuts, berries, pulp fruits and leaves of these trees, and probably the occasional insect. A few are particularly fond of palm oil nuts, which they peel and eat in their dens.

Like beavers, the scaly-tailed squirrels use their gnawing incisors not only to eat bark, but also to cut branches from trees. Although this behavior is not fully studied, what we know so far suggests that scaly-tailed squirrels use tree pruning to manage their environment in just as

SCALY-TAILED SQUIRRELS

CLASS **Mammalia**

ORDER **Rodentia**

FAMILY **Anomaluridae**

GENERA AND SPECIES **Beecroft's scaly-tailed squirrel, *Anomalurus beecrofti*; pygmy scaly-tailed squirrel, *Idiurus zenkeri*; flightless scaly-tailed squirrel, *Zenkerella insignis*; others**

ALTERNATIVE NAMES
Anomalure, scalytail, scaly-tailed flying squirrel, flying mouse (*Idiurus* spp.)

WEIGHT
⅖ – 69 oz. (14–2,000 g)

LENGTH
Head and body: 2⅖–17 in. (6–43 cm); tail: 2⅖–18 in. (6–46 cm)

DISTINCTIVE FEATURES
Flight membrane supported by rod of cartilage emerging from front elbows; two rows of overlapping scales at base of underside of tail

DIET
Bark, fruit, flowers, leaves, nuts, insects

BREEDING
Age at first breeding: not known; breeding season: probably year round; number of young: probably 1 to 4; gestation period: not known; breeding interval: not known, but probably 6 months minimum

LIFE SPAN
Not known

HABITAT
Canopy of African tropical and subtropical forest

DISTRIBUTION
Sub-Saharan Africa, within Tropics

STATUS
Near threatened

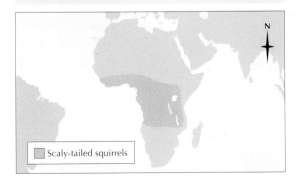

Scaly-tailed squirrels

---

ingenious a way as do beavers. One function of pruning is to clear flight paths underneath the canopy, from tree to tree. Pruning subcanopy branches also encourages dieback, a defensive response of the tree which, in some species, creates a shallow hole in the main trunk. The scaly-tailed squirrel can then enlarge this hole into a den for itself. The animals also cut the tops off young trees in the vicinity of their food trees. The maiming or death of the food trees' competitors must benefit the food trees, and it is possible that scaly-tailed squirrels and their food trees live in an intimate, mutually beneficial partnership. At least one food tree, the awoura, *Julbernardia pellegriniana*, can survive the complete removal of its bark by scaly-tailed squirrels, and can regrow it.

Fighting for trees

Scaly-tailed squirrels usually live in pairs or family parties but they may form larger groups; the pygmy scaly-tailed squirrels sometimes roost in large numbers. Social life and breeding is conducted in the canopy and is difficult to study, but it seems the females have one, two or sometimes four babies in each litter, and there may be two litters a year.

The scaly-tailed squirrels must compete for their dens in tree holes with other animals, especially hornbills. Their favorite food trees are also prized timber species, so they are bound to come increasingly into conflict with humans. Most scaly-tailed squirrel species are rare and little known. We do not know how vulnerable they are, or how precariously dependent they might be on specific habitats, so more study is urgently needed.

Pel's anomalure is entirely nocturnal. It communicates vocally with deep hoots. It is large and aggressive, and its bold coloring may indicate it has less to fear from predators than other scaly-tailed squirrels.

SCARAB BEETLE

THE MOST FAMOUS scarab beetle is the sacred scarab, *Scarabaeus sacer*, of Egypt. The scarab and its immediate relatives are scavengers, and many of them are dung beetles. They collect and bury dung in a variety of ways, but the best-known method is to roll a dung ball along the ground. This has led to the name of tumblebug in the United States. The sacred scarab is only one of more than 20,000 species in the family. Others include the cockchafers (discussed elsewhere) and the largest beetles in the world: the goliath beetle, 4 inches (10 cm) long; the equally large rhinoceros beetle; and the hercules beetle. The scarabs and dung beetles are mostly medium sized, ½–1 inch (1.3–2.5 cm) or longer, with heavy bodies, oval or almost oblong in outline, and a bumbling flight. Many are colored in metallic hues of black, purple, blue, green, bronze or gold, usually iridescent. Their legs are short and the middle joints are flattened and broadened. The antennae are short but the head and thorax often bear spines or spikes.

A green scarab beetle, Kheper aegyptiorum, with a ball of buffalo dung. Some scarab beetles lay their eggs in the dung where it lies; others carefully roll it away for burial.

There are dung beetles in every continent except Antarctica, but the number of species in each continent varies. In Australia and South America, where there are fewer large mammals than in Africa, the number of species of scarab beetles is proportionately smaller.

Large-scale food transport

The typical scarab, on finding a pile of dung, separates a portion of it, compresses it and molds it into a ball much larger than itself and then starts to roll it away. An inch-long beetle may make a ball 3–4 inches (7.5–10 cm) in diameter. The beetle places its hind legs on the ball and, with its body inclined downward and front feet pushing on the ground, proceeds to roll the ball along, perhaps helped by another of its species. Having taken the load a distance, the beetles dig a hole under it until it drops down and is buried. They feed on this ball, eating more than their own weight in a day, and then return to the surface for a fresh supply.

SCARAB BEETLES

PHYLUM	**Arthropoda**
CLASS	**Insecta**
ORDER	**Coleoptera**
FAMILY	**Scarabaeidae**
GENUS AND SPECIES	**About 20,000 species worldwide**

ALTERNATIVE NAMES
Dung beetles; chafers; hercules beetles; goliath beetles; many others

LENGTH
From less than ⅕ in. (5 mm) to more than 7 in. (18 cm)

DISTINCTIVE FEATURES
Extremely varied; some have large hornlike structures on head or thorax; metallic sheen; antennae usually short, ending in a club

DIET
Scarab larvae: dung. Chafer larvae: plant material. Others: decaying matter or fungi.

BREEDING
Extremely varied; parental care also variable

LIFE SPAN
Several years for some species

HABITAT
Extremely varied. Dung beetles: wherever preferred dung found; underground tunnels containing dung collected from surface. Chafers: wherever host plants found. Larvae: generally underground.

DISTRIBUTION
Worldwide

STATUS
Common

A silver striped scarab beetle, Plusiotus gloriosa, *on the arid grassland of Arizona. The scarabs include some of the largest beetles in the world, and many smaller ones.*

Not all dung beetles are ball-rollers. As might be expected in such a diverse group, the method of feeding is very varied. Some species eat and lay eggs in the dung where they find it, others roll balls and transport them. Some bury the dung while others tunnel into cowpats.

Taking care of the young

When they are about to breed, the scarabs come together in pairs and the pattern of the dung collecting changes. The two combine to roll a large ball to a selected site, where they bury it. When it is underground, the female shreds the dung and molds it into a solid pear-shaped mass, tamping its surface hard. The neck of this mass is left hollow and an egg is laid in the cavity. A hole is left at the top, which is loosely filled so as to allow air in. The adults leave the cavity, filling in the tunnel behind them as they go up to the surface. When the larva hatches, it feeds on the dung and then pupates. When rain softens the earth above, the new beetle emerges from the pupal skin and makes its way to the surface. This behavior is typical, but the details vary between species of scarab beetles.

The larva has complete protection in its underground larder, so only a few eggs, usually two to four, are laid in one season. Other species of dung beetles use less elaborate methods, burying smaller quantities of dung in egg-shaped, globular masses or in cylinders. The dor-beetle, genus *Geotrupes*, digs itself a vertical shaft 8 inches (20 cm) deep with a plug of dung at the end. All these species lay more eggs. In a European species, *Copris hispanus*, the female stays underground, constantly walking around the several balls of dung she and her mate have accumulated in one cavity. Each ball contains one egg, and the mother works incessantly to keep the surfaces of the balls clear of molds, sealing any cracks that may develop in them. This is because they have not been tamped and hardened as in the sacred scarab and others. Indian scarabs of the genera *Heliocopris* and *Catharsius*, for example, coat the large balls of dung with clay and bury them 8 feet (2.4 m) down. However, parental devotion ceases as soon as all the young beetles have hatched and have found their way to the surface.

The larvae of many scarabs and related dung beetles stridulate (issue high-pitched sounds by rubbing body parts together) and the adults of some species are also known to make similar sounds. The purpose of this is not clear.

Parasites and predators

G. stercorarius is known in Britain as the lousy watchman because it is so frequently infested with mites. These mites push their sucking mouthparts through the soft skin at the joints of the beetle's armor and suck the blood. Predators include shrews, hedgehogs and similar animals. Within the last few years scientists have discovered that hibernating horseshoe bats wake at repeated intervals throughout the winter and fly out to feed on dung beetles.

An ancient deity

The ancient Egyptians frequently depicted the scarab on their monuments, bas-reliefs, jewelry and seals. Historians believe that they assumed a gigantic celestial scarab kept the Earth revolving just as the scarab made its ball of dung revolve. On the front of the sacred scarab the head is drawn out into a number of spines, which to the Egyptians symbolized the sun's rays, giving the beetle an even greater significance. They also believed the beetle buried its ball of dung for 29 days and on the thirtieth day cast it into the Nile, rounding off the month. There could have been other, more practical reasons for deifying the scarab, such as its all-important task of sanitation, while at the same time making the ground more fertile.

Scarab beetles often collect a dung ball many times their own size and roll it along using their hind legs.

SCAT

SCATS ARE SO VARIABLE in color that it is difficult to know the exact number of species, but scientists believe there are four in total. They are disk-shaped fish with small heads and bodies that are flattened from side to side and may grow to 15 inches (38 cm) long, although most are usually 3–5 inches (7.5–12.5 cm). Both dorsal and anal fins are made up of a spiny portion in front and a soft-rayed part behind, the two parts being more or less continuous. When scats accelerate through the water they hold their spines flat against the body, but when they turn suddenly, or slow down quickly, the spines are raised. The pectoral and pelvic fins are small, the pelvic fins lie far forward on a level with the pectorals and the tail fin is relatively large and square ended. The body and the bases of the fins are covered with small scales.

The background color varies according to a number of criteria, including the species, the individual, and age, but a fairly constant feature is the rows of large spots, often running together and forming bands, like the stripes of a tiger. Indeed, one variety is known as the tiger scat,

Scatophagus rubifrons. Scats are also sometimes referred to as argusfish, because their many spots recall the monster from Greek mythology called the argus, which was reputed to have 100 eyes. Nineteenth-century zoologists were often classical scholars as well as scientists and the scientific names that they used to classify animals frequently reflected this fact.

The best-known scat is the spotted scat, *S. argus*, which, when 2 inches (5 cm) long, is greenish silver, bluish silver or coffee brown, with rows of large dark spots or incomplete bars. As the fish gets older, the back becomes reddish or orange, and this color later spreads down the flanks, so the fish appears irregularly barred with orange and black. Other individual fish may be uniformly colored and marked with black spots.

Scats are found in the Indian Ocean and the South Pacific, from Kuwait to Vanuatu and New Caledonia, north to southern Japan, and from Palau to Pohnpei in Micronesia. One species, called the scatty, *S. tetracanthus*, is found in East African waters as well as around northern Australia and New Guinea.

The dorsal, anal and pelvic spines of scats, such as the spotted scat (above), are believed by Philippine fishers to be venomous and capable of inflicting wounds.

Scats are coastal fish but they also frequent sheltered bays, estuaries and the lower reaches of streams. They are often to be found in mangrove swamps.

Reef-living browsers

All scats are coastal fish that come into estuaries and fresh waters. They remain there until they are well grown, when they go back to the sea. Scats are attractive, if expensive, aquarium fish and are more or less constantly active. They swim with the unsteady movement typical of coral fish, which is one reason for supposing they are essentially reef living. Another reason for this supposition is that they browse on almost any plant in an aquarium, eating it down to its roots. Scats have an almost insatiable appetite and eat many types of plant food, as well as dead animals. They are reputed to feed on mud, decomposing refuse and even human excrement, and they congregate in large numbers around sewer effluents. One of the first scientists to examine the stomach contents of these fish found only offal, so the fish were given the generic name *Scatophagus*, meaning dung-eater.

Breeding

There is still some scientific uncertainty about the breeding habits of scats, although *S. tetracanthus* can reproduce in fresh water. There seems to be no easy means of distinguishing the male from the female. Scats have only once been bred in captivity. The eggs were laid in crevices in rocks but the newly hatched larvae did not survive. It is thought that when they leave the rivers and estuaries for the sea, they go there to breed, possibly on the coral reefs. In aquaria there are indications that both male and female scats tend the eggs and larvae. The larvae pass through a stage in which the head and nape are covered in a bony armor, but this is lost before the fish becomes an adult.

CLASS	**Osteichthyes**
ORDER	**Perciformes**
FAMILY	**Scatophagidae**
GENUS AND SPECIES	***Scatophagus argus***

LENGTH
Up to 15 in. (38 cm)

DISTINCTIVE FEATURES
Deep, strongly compressed, disk-shaped body; distinct lateral line; small head and mouth; variable coloration, greenish silver, bluish silver, coffee brown; golden sheen, especially on back; transverse bars; large, round blackish spots or bars; wide variety in coloration of fins

DIET
Worms, crustaceans, insects and plant matter

BREEDING
Little known; spawns in rock crevices

LIFE SPAN
Not known

HABITAT
Harbors, bays, brackish estuaries; lower reaches of freshwater streams; often occurs among mangroves

DISTRIBUTION
Indian Ocean and South Pacific: Kuwait southeast to Vanuatu and New Caledonia; north to southern Japan; Palau east to Pohnpei in Micronesia; Samoa; Society Islands, French Polynesia

STATUS
Common

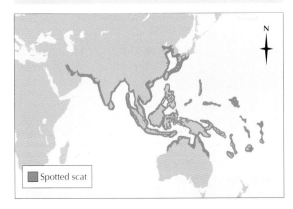

Spotted scat

SCORPION

CORPIONS ARE PROBABLY best known for their stings, the venom of which, in some species, is fatal to humans. They range in length from ¼ inch (6 mm) to as much as 8 inches (20 cm) in the emperor scorpion, *Pandinus imperator*. The pedipalps, appendages used in sensing or feeding in other arachnids such as spiders, are modified and enlarged in scorpions into large, pincer-like claws. These claws might be similar in appearance to those of a lobster, but lobster claws are formed from chelicerae, which are mouthparts. Other groups of arachnids also show pedipalps enlarged into claws, including the pseudoscorpions.

The body of a scorpion is segmented. The thorax has four segments, each with a pair of walking legs on the undersurface. The abdomen has six segments tapering to a sharp stinger, or telson, at the tip. The stinger has a small opening supplied by two relatively large venom glands. There are about 40 North American scorpion species, but only two, the 2–3-inch (5.0–7.5-cm) *Centruroides sculpturatus* and *C. gertschi*, have enough venom to be lethal to humans. Those scorpion species that are dangerous to humans are restricted largely to the family Buthidae.

There are about 1,500 species of scorpions altogether. Most are found in warmer climates, although some live in relatively cool, moist environments. In North America scorpions are found as far north as British Columbia, and they also live in southern Europe. Scorpions are extremely adaptable and many species can withstand the fierce heat of desert climates.

In some regions scorpions may be found in human habitations. Away from houses, they hide by day under logs or rocks or in holes in the sand, which they dig with their middle legs. Scorpions tend to lead solitary lives, unless they are engaged in courtship or rearing of young. They are often hostile to other scorpions; the females may even devour the males after mating.

Tearing their victims to pieces

Most species of scorpions are nocturnal animals. Their prey consists almost entirely of insects and spiders and they usually seize their victim with their large claws and tear it to pieces, or crush it, before starting to feed. They might eat all parts of the prey, or just suck out the soft parts. The scorpion may use its venom to subdue the prey, especially if the victim offers any resistance. The

The scorpion **Buthus occitanus** *of Spain is a member of the* **Buthidae,** *the family that contains most of the highly venomous species of scorpions.*

The Asian forest scorpions of the genus Heterometrus *rival the West African emperor scorpion in terms of size. Their sting is painful, but not medically significant.*

SCORPIONS

PHYLUM	**Arthropoda**
CLASS	**Arachnida**
ORDER	**Scorpiones**
FAMILY	**Several, including Bothriuridae, Buthidae, Chacticae and Scorpionidae**
GENUS	**Many**
SPECIES	**About 1,500**

LENGTH
⅖–8 in. (1–20 cm)

DISTINGUISHING FEATURES
Flattened body; well-developed pincers at front; abdomen narrowed at end and tipped with sting; four pairs of walking legs

DIET
Other arthropods, mainly insects or arachnids

BREEDING
Following courtship dance, male pulls female over spermatophore on ground. Female takes up spermatophore with her cloaca. All scorpions either ovoviviparous or viviparous.

LIFE SPAN
Up to several years in some species

HABITAT
Cracks and crevices under stones, among rocks and plants, usually in warmer areas of deserts, caves, forests and other habitats

DISTRIBUTION
All continents except Antarctica

STATUS
Common

scorpion brings its abdomen forward over the body and thrusts the venom-bearing stinger into the prey. The prey is then slowly eaten, an hour or more sometimes being spent in consuming a single beetle.

Scorpions can survive long periods without eating and it is said that they never drink, getting all the moisture they need from their food or from dew. In captivity, however, they regularly drink water. It can be supplied to them using a wet wad of cotton wool.

Some scorpions stridulate, or sing, by rubbing the bases of their clawed limbs against the bases of the first pair of legs. In some species, the sound is generated by a rasp at the base of each claw. Scorpions do not stridulate for the same reason grasshoppers do. Their song is used to announce their intention to attack another scorpion, or as a defensive warning. The positions of the scorpions' claws differ according to their intention: in the attack posture the claws are held open, wide apart and pointing upward, whereas in the defensive position the claws are held low and in front of the head.

Courtship dance

Like spiders, scorpions go through a form of courtship before mating. Normally, when the female shows herself to be responsive, the male first grasps her with his claws and then maneuvers to face her, gripping her claws with his own; the courtship of some scorpion species may also involve stinging. Sometimes, when the female is not submissive and tries to pull away, the male raises his stinger almost straight above his claws and the female does the same. Having grasped the female, the male drags or pushes her to a suitable place, where he scrapes away the soil with his feet and deposits his spermatophore (package of sperm). Then, still holding his partner by her claws, he maneuvers her over it so she can take up the spermatophore with her cloaca (a chamber into which the urinary, intestinal and generative canals feed). The two animals remain together for 5 or 6 minutes before breaking away.

Unlike spiders, scorpions do not lay eggs. Development in scorpions instead takes one of two forms, both unusual in arthropods: ovoviviparity and viviparity. In ovoviviparous species, the embryos develop inside eggs within the mother and are born as hatched young. In viviparous species, embryos also develop inside the mother, but are not separated by an egg membrane, and are nourished directly by her tissues. In these cases, a placentalike tissue may be formed in the mother's body through which nutrients are passed to the growing embryo and waste products are removed.

The young scorpions are born one or two at a time over a long period of up to several weeks, depending on species and environmental conditions. Initially, newborn scorpions ride around on their mother's back and only become independent from their mother after their first molt.

Dangerous when provoked

People are most at risk from scorpion attacks when the animals enter houses. Scorpions will not use their sting against humans unless they feel threatened or provoked, but where they live in close proximity to humans, it is easy for conflicts to arise accidentally. This problem can become significant where scorpions are common. In the United States and Mexico it is estimated that more people are killed by scorpions than by snakes. In 1954, nearly 200 people needed emergency hospital treatment for scorpion stings in a Brazilian town with a population of only 200,000. Similar reports may be found in other tropical parts of the world.

A sting from one of the more dangerous scorpion species is often followed by collapse, profuse perspiring and vomiting, and the skin becoming cold and clammy. Even when drugs and oxygen are administered the patient may have difficulty breathing, and in the worst cases the sting may prove fatal. However, there is considerable variation in the potency of a scorpion's sting, according to species, and the great majority cause as little reaction as a bee sting.

Young scorpions ride on their mother's back. Scorpions do not have an external egg stage, and the young emerge as well-developed juveniles. The species pictured lives in the dry forests of Costa Rica.

SCORPIONFISH

The devil scorpionfish, Scorpaenopsis diabolus, feeds on diurnal (day-active) reef fish. In this photograph the victim is a butterflyfish.

THERE ARE SOME 300 species of scorpionfish. Most live in temperate seas, although a few are found in the Tropics. Scientists originally classified 388 species of scorpionfish, grouped into 52 genera, within the family Scorpaenidae. However, after subsequent adjustments in taxonomic criteria, certain subfamilies are now regarded as families in their own right. The family Scorpaenidae now consists of 172 species in 23 genera. These comprise some of the most poisonous fish in the world.

Native to the Indian Ocean and South Pacific, the stonefish, *Synanceia horrida*, is known for its lethally venomous spines. The red lionfish, *Pterois volitans*, also known as the turkeyfish, lionfish, zebrafish, dragonfish or butterfly cod, is one of the most visually arresting scorpionfish. The body, which grows up to 12 inches (30 cm) or more long, has striped, zebralike markings. The massive head is irregular in shape, with eyes set high up and the mouth wide and sloping down at the corners. All the fins are divided into ribbonlike strips, those of the pectoral fins being longer than the body. Among these are poison spines, 13 in the dorsal fin, 3 in the anal fin and 1 in each pelvic fin. The colors may be maroon with gray stripes, rose with bluish-white stripes or brown with yellow stripes.

Another member of the family is the California scorpionfish, *Scorpaena guttata*, also known as the sculpin, a name that is more accurately associated with members of the bullhead family.

The sculpin has the large head of a bullhead flanked by sizeable, rounded pectoral fins and may be 12 inches (30 cm) or more long. Its body is marked with warts, flaps and frills, which, with its mottled color, make it hard to see among the seaweeds on which it lies. The barbfish, *S. brasiliensis*, is an even more ornate species, featuring flaps, frills, tiny hooks and barbs, as well as patches of color. It has the same shape as the California scorpionfish but shows nearly all the colors of the rainbow, from blue patches on the head, red around the mouth and throat, to a motley of reds, yellows, browns and purples on the body and fins. The orange scorpionfish, *S. scrofa*, which is less colorful but which has the same shape, is a brownish orange. In the bearded roguefish, *Tetrarope barbata*, the dorsal fin is more like a sail than a row of spines, and its colors are somber browns, yellow and orange. The Norway redfish, *Sebastes viviparus*, grows up to 3 feet (90 cm) long, is reddish in color and lacks the flamboyant fins of the other species in the family Scorpaenidae. It has two near relatives: the ocean perch, *Sebastes marinus*, and the blackbelly rosefish, *Heliocolenus dactylopterus*, also known as the bluemouth, or bluethroat.

The California scorpionfish lives off the Pacific Coast of North America, while the barbfish is found off the Atlantic Coast of tropical America, from New Jersey south to Rio de Janeiro. The bearded roguefish lives around the Philippines, the Norway redfish lives on both sides of the North Atlantic and the ocean perch and blackbelly rosefish live in the eastern North Atlantic.

Poison spines

Scorpionfish range from shallow waters to deep waters. In general, the deeper the water in which they live, the more they tend toward a single color. The shallower the water, the more broken up and varied are the colors of the body, the more elaborate the fins and the more irregular the body surface. The shallow-water members of the family are also more poisonous, but none is as poisonous as the stonefish.

There is another difference between the stonefish and other members of the family. The stonefish raises its spines and remains perfectly still when it is approached; the red lionfish and others raise their spines and move them, and also

SCORPIONFISH

CLASS	**Osteichthyes**
ORDER	**Scorpaeniformes**
FAMILY	**Scorpaenidae**
GENUS	**23 genera**
SPECIES	**172, including blackbelly rosefish,** *Helicolenus dactylopterus* **(detailed below)**

ALTERNATIVE NAMES
Bluemouth; bluethroat

LENGTH
Up to 18 in. (45 cm)

DISTINCTIVE FEATURES
Relatively narrow-bodied fish; moderately large, spiny head; mainly red body, becoming rose pink ventrally; inside of mouth and gill cavity dark leaden blue

DIET
Mainly small fish and crustaceans such as shrimps and prawns; also small squid, octopuses, brittlestars and starfish

BREEDING
Oviparous. Breeding season: mainly November–December.

LIFE SPAN
Up to 25 years

HABITAT
Soft-bottom areas of continental shelf and upper slope at depths of 660–3,280 ft. (200–1,000 m)

DISTRIBUTION
Western Atlantic: Nova Scotia south to Venezuela. Eastern Atlantic: Iceland and Norway south to Mediterranean and Gulf of Guinea, including Madeira, the Azores and the Canaries; Walvis Bay, Namibia, south to Natal.

STATUS
Not threatened

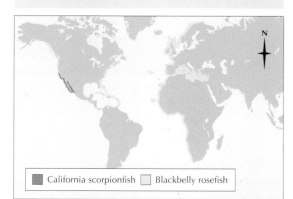

☐ California scorpionfish ☐ Blackbelly rosefish

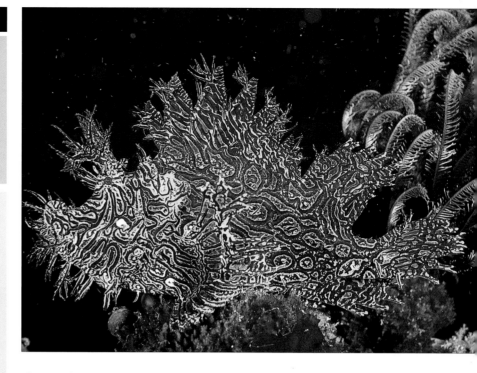

The weedy scorpionfish, Rhinopias aphanes, enhances its superb camouflage by resting next to featherstars.

change the position of the body to present their spines to the best advantage in an attack. They may even jab at the intruder with their spines. Most of them give painful but not fatal wounds. The pain from a California scorpionfish, for example, lasts an hour, though the swelling may last much longer. Despite their poisonous spines and unappetizing appearance, the flesh of scorpionfish is palatable. However, they are not fished commercially.

Small fish the target

In all scorpionfish the poison spines are used only for defense, not for killing prey. Scorpionfish feed mainly on smaller fish. Those living in shallow waters usually lie concealed on the bottom, snapping up small fish that swim past their mouths. They also take shrimps, prawns and other crustaceans. Those scorpionfish, such as the Norway redfish and the related ocean perch, that swim actively feed on a wider variety of small crustaceans as well as fish, but as they grow larger they eat a greater proportion of fish.

Balloons of spawn

Most scorpionfish give birth to free-swimming young, the eggs hatching inside the female's body. There are reports of single females giving birth to thousands of young at a time. The California scorpionfish, one of the few members of the family to lay eggs, spawns several times during the summer. The eggs are laid embedded in two hollow, pear-shaped balloons of jelly, joined at their small ends. These float at the surface, but the newborn fish sink to the seabed as they hatch.

SCREAMER

The screamer's long legs and long, mostly unwebbed toes enable it to walk easily among floating vegetation. Pictured is the southern screamer.

SCREAMERS ARE RELATED to ducks, geese and swans. All belong to the order Anseriformes, are aquatic and live in flocks. Although heavy-bodied like ducks and geese, screamers have longer legs with only small webs between the toes, and the bill is hooked and chickenlike. The hind toe is long and on the same level as the three front toes. The wings bear two sharp spurs, one 1–2 inches (2.5–5 cm) long, and also a blunt boss (protuberance).

The southern screamer, *Chauna torquata*, is the best known of the three screamer species. It is about the size of a goose and has gray plumage with a black ruff around the neck and a short crest. It ranges from eastern Bolivia and southern Brazil to central Argentina. The northern screamer, *C. chavaria*, has much darker plumage, a black neck with a white ring at its base, white cheeks and a longer crest. It is found in a small area of northern Colombia and northwestern Venezuela. The horned screamer, *Anhima cornuta*, ranges from Venezuela to eastern Bolivia and southern Brazil, and once also lived on Trinidad. This bird is a little larger than the northern screamer and is about the same size as the southern screamer. Its plumage is mainly glossy greenish black with a white belly. A narrow, stiff, cartilaginous horn curves forward from the forehead.

Two-note trumpet

Screamers live in areas of marshes, lagoons or damp grassland, and sometimes also feed among grazing livestock. In marshes and lagoons they prefer the masses of floating vegetation, in which they can walk about supported by their long toes. The horned screamer is also found along slow streams. Screamers rarely swim unless forced to. When they do so, they swim well, although they move slowly and hold themselves high in the water. Their plumage appears to wet easily and they frequently hold their wings out to dry them. However, they fly well, rising heavily but flying easily once airborne and often soaring so high that it is difficult to identify them.

Screamers are so called because they have a very loud, two-note, trumpeting call. During the breeding season the pairs call to each other, and outside the breeding season screamers gather in flocks of several hundreds and scream together. They have a second call that is a low rumble, audible only at close quarters.

Screamers are plant-eaters, cropping grass and marsh plants. The horned and southern screamers also eat insects, a food that the chicks of these species particularly favor.

Violent courtship

Pairing starts in early spring, each pair forming a territory. There is a certain amount of fighting that may lead to severe injury. The screamer sheds the outer layers of its wing spur periodically. These have been found in the breasts of many screamers, suggesting that the spurs are used in aggressive confrontations. The breeding habits have not been studied in detail, but scientists have discovered that the southern screamer tramples an area of plants in shallow water close to the shore and builds its nest of plant matter in the middle of this flattened area. The two to seven white eggs are incubated by each parent in turn for 6–7 weeks. If it is disturbed, the sitting bird does not leave the nest until the very last moment, when the intruder is only a short distance away. When it does leave the nest, it tries to frighten the intruder by flapping its wings and sometimes slashing with the spurs on its wings. The chicks, which somewhat resemble goslings, leave the nest soon after hatching and follow their parents.

Internal air sacs

A unique feature of the respiratory system of birds is the air sacs that lead off from the lungs. These are thin-walled bags that lie in the body and even pass into the bones, such as those in the wing. The precise function of the air sacs is not fully understood by scientists, but it is thought that they must increase the efficiency of breathing by allowing air to circulate through the lungs during both inspiration and expiration. This would allow more oxygen to pass from the air into the blood via a kind of double-pump action by which the fullest use is made of oxygen in the air. In flight a bird uses up oxygen in the same way that an athlete does when running, but does so without panting. Also, the presence of air sacs in the bones makes them relatively lighter than those of terrestrial animals.

Screamers are unusual among birds because of the extent of their air sacs, which not only form bubbles under the skin but also reach as far as the toes. When these sacs are pressed, they make a crackling sound similar to the popping of tiny balloons. In addition to their probable function in respiration, the air sacs may also be in some way connected with the low rumbling noises that the birds make.

NORTHERN SCREAMER

CLASS	**Aves**
ORDER	**Anseriformes**
FAMILY	**Anhimidae**
GENUS AND SPECIES	***Chauna chavaria***

LENGTH
Head to tail: 30–36½ in. (76–91 cm)

DISTINCTIVE FEATURES
Hooked, chickenlike bill; longer legs than a duck or goose; 2 sharp spurs on each wing; reduced webbing on feet; gray crown and crest; rest of head white; black neck; remainder of plumage dark gray, glossed green above; white underwing coverts

DIET
Roots, leaves, stems and other green parts of succulent aquatic plants

BREEDING
Breeding season: most eggs laid October–November; number of eggs: 2 to 7, usually 3 to 5; incubation period: 42–44 days

LIFE SPAN
Not known

HABITAT
Swamps, marshes, lagoons, banks of slow-flowing rivers and seasonally flooded alluvial plains

DISTRIBUTION
Northern Colombia and Venezuela

STATUS
Near threatened, but common at some localities within its restricted range

Northern screamer

Despite their heavy bodies, screamers are strong fliers and are able to reach great heights. They sometimes perch in trees, particularly at any sign of danger.

SEABIRDS

Several groups of birds spend all, or much, of their lives at sea. Some are truly oceanic, traveling far from land for months at a time, while others stay in coastal waters. For convenience, ornithologists often refer to all of these species as seabirds, though the term has no precise meaning. This guidepost article deals with the members of three different orders: Charadriiformes (including the jaegers, terns, gulls, skimmers and auks), Pelecaniformes (frigate birds, tropic birds, gannets, cormorants and pelicans) and Procellariiformes (petrels, albatrosses, storm petrels and diving petrels). The birds that belong to these orders are varied in both their appearance and their habits. The penguins, which belong to the order Sphenisciformes, are also seabirds but are discussed in a separate guidepost article.

Many species of birds are adapted for life above and within the world's oceans, in estuaries and offshore shallows, in coastal seas and ranging across the oceans, hundreds of miles from land. Seabirds are found in a wide range of sizes, from the small auks, prions and diving petrels to the massive pelicans, which may be up to 17½ pounds (8 kg) in weight.

Adaptable species

A diverse range of birds is adapted to life at sea. Shorelines and mudflats are inhabited by shorebirds, geese and ducks, which dabble and feed in the mud. In shallow coastal waters, they are joined by fish-eating birds such as loons, grebes and sawbills. The flightless penguins range across the seas of the Southern Hemisphere.

Many seabirds, including the Atlantic puffin, Fratercula arctica *(above), choose inhospitable locations, such as sea cliffs, to nest. By doing so they make themselves and their young less vulnerable to attack from foxes, rats and other predators.*

A wide-ranging distribution

Almost all seabirds spend only the breeding season on or close to land, living at sea the rest of the year. Some, such as the auks, spend this time in shallow waters a few miles offshore, often on the surface of the sea; others fly across vast stretches of ocean, far from land. These pelagic (oceanic) species include the shearwaters and albatrosses. The short-tailed shearwater, *Puffinus tenuirostris*, circumnavigates the Pacific Ocean before returning to breeding grounds on small islands off the coast of Australia. Albatrosses spend most of the year flying low over the waters of the Southern Ocean.

Many seabirds migrate with the changing seasons. Some, including the gulls, may migrate only a small distance, or perhaps move inland. Others cover longer stretches, usually to warmer climates. For example, the Cape gannet, *Sula capensis*, moves north along the coast of Africa during the fall, while the closely related northern gannet, *S. bassana*, moves in the opposite direction, wintering off the North African coast or the Gulf of Mexico. Forster's tern, *Sterna forsteri*, migrates from southern Canada as far south as Venezuela. However, a few seabirds migrate to cooler regions after the breeding season is over. Xantus' murrelet, *Endomychura hypoleuca*, breeds in California but flies as far north as Washington in the winter. By far the longest migration is that of the Arctic tern, *Sterna paradisaea*. Breeding just south of the Arctic Circle, it spends the rest of the year flying south to the Antarctic and back, with the benefit of almost constant summer weather as it flies.

Most seabirds are solitary and come together only during the breeding season. They may, however, gather together in groups at a rich food source, such as the offal from a fishing ship.

Ways of feeding

The sea is a rich, productive environment, and a relatively small area can support many different species of birds, often over a range of sizes. Most of the larger seabirds specialize in fish and squid, but many, such as terns, also eat small shrimps, while little auks or dovekies, *Alle alle*, are plankton feeders. Some of the more widespread seabirds, such as the fulmars and gulls, are generalist feeders and eat nearly any form of food. Fish and squid eaters avoid competition by specializing on prey living at different depths and of different sizes. Auks select fish on the basis of diameter, rather than length or overall size. The gape of the bill determines which fish are caught, allowing razorbills, puffins and murres to feed noncompetitively in the same areas, because their bill gapes are different, and the birds therefore capture different-sized fish.

Seabirds use a variety of fishing methods. Many birds, including gannets, tropic birds and terns, are plunge divers. These birds swoop down steeply into the water, diving into the midst of a shoal of fish. Brown pelicans, *Pelecanus occidentalis*, also feed in this manner, scooping fish into an elastic throat pouch, which acts as a net. Some pelicans work together to drive fish into shallower water. Pelicans generally swallow their prey at the surface, while gannets swallow theirs underwater, on the way back to the surface. Plunge dives are usually short, lasting 5 seconds or less.

Plunge divers do not usually swim underwater in pursuit of their prey. By contrast, birds such as cormorants dive from the surface and pursue their prey underwater with powerful thrusts from their webbed feet. Captured fish are returned to the surface, where they are swallowed whole. The diving petrels and auks demonstrate an alternative method of underwater pursuit. These birds pursue fish and squid by actively swimming underwater using their wings. They can dive more deeply than most plunge divers, and the largest living auks are able to stay submerged for up to 3 minutes before having to return to the surface. Albatrosses feed at the surface of the water, taking mainly shrimps and small fish. They also feed on dead or dying marine organisms, as many gulls do, and these birds sometimes eat the rubbish strewn behind fishing ships.

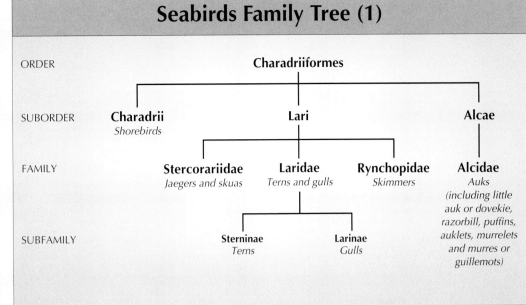

Seabirds Family Tree (1)

ORDER		**Charadriiformes**		
SUBORDER	**Charadrii** *Shorebirds*	**Lari**		**Alcae**
FAMILY		**Stercorariidae** *Jaegers and skuas* **Laridae** *Terns and gulls* **Rynchopidae** *Skimmers*		**Alcidae** *Auks (including little auk or dovekie, razorbill, puffins, auklets, murrelets and murres or guillemots)*
SUBFAMILY		**Sterninae** *Terns* **Larinae** *Gulls*		

Like other cormorants, spotted shags, Stictocarbo punctatus, breed in large colonies. The male chooses the nest site and collects material for the nest, which the female builds.

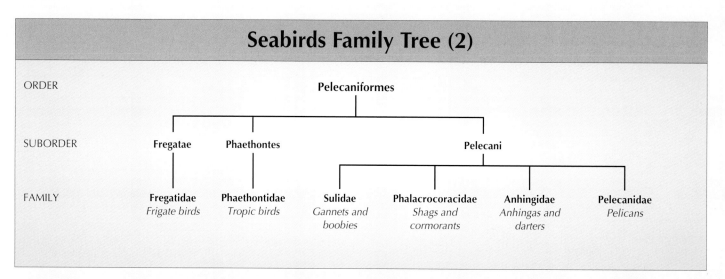

Seabirds Family Tree (2)

ORDER			Pelecaniformes			
SUBORDER	Fregatae	Phaethontes		Pelecani		
FAMILY	**Fregatidae** *Frigate birds*	**Phaethontidae** *Tropic birds*	**Sulidae** *Gannets and boobies*	**Phalacrocoracidae** *Shags and cormorants*	**Anhingidae** *Anhingas and darters*	**Pelecanidae** *Pelicans*

Skimmers employ a remarkable method of feeding. They fly low over the water, trailing their elongated lower bill in the swell. When it comes into contact with a small fish or shrimp, the much shorter upper bill clamps down. These birds fish at dusk, when their prey comes close to the surface to feed. Petrels also fly low over the water, often with a characteristic fluttering flight path. Some of these birds have brightly colored webbed feet. This may act as a lure for the plankton that form their prey, as the birds often paddle in the water as they fly. Storm petrels are sometimes found with missing legs; it is possible that the colored webbing also attracts larger, predatory fish, which nip at the birds' feet. They may also lose limbs when attacked by skuas and jaegers.

Frigate birds have very large wings, but they are relatively light, making them highly maneuverable in flight. They exploit this by harrying and harassing smaller or less agile seabirds as they return from the fishing grounds to the nest. Often the bird either drops its load of fish, or regurgitates food from the crop; this lightening of the load allows the bird to escape, but provides the frigate bird with an easy meal. Frigate birds are uneasy on the sea, and rarely alight on it;

Gulls have a global distribution, although most favor northern temperate regions. Many gull species are found in very harsh environments. For example, Ross's gull, Rhodostethia rosea *(below), breeds almost exclusively in northeastern Siberia.*

they steal most of their food from other birds. Skuas and jaegers also steal food from auks and terns; they also occasionally feed upon the birds themselves, or their young.

Traveling by air and sea

As with most birds, wing shape is an important determinant of ecological niche (the ecological role of an organism in a community). Auks and diving petrels use their wings both for swimming and for flight; the wings of these birds are relatively small, so that the birds can use them to swim in the water. However, their wing size means that in the air these birds are rather poor at slow, maneuverable flight, and they must beat their wings faster than most other birds to stay in the air. Flying is an energetically costly business, and for birds such as these it can prove more efficient to dispense with flight altogether and rely solely on swimming to travel. Penguins are one example of birds that employ this method of travel today, but from the Miocene (part of the Tertiary geologic time period, between 37 and 24 million years ago) until the early 19th century, northern waters abounded with a range of flightless auks.

Birds with long, pointed wings, such as terns, are efficient fliers and are well adapted for long migratory flights. Many seabirds are

A pair of wandering albatrosses, Diomedea exulans, *extends its wings during display. This species' wingspan, which may reach 11 feet (3.4 m), is the largest of any living bird.*

skilled gliders. Some, such as gulls and fulmars, use the air currents that rise where the sea meets a cliff to wheel and soar. The most accomplished practitioners of gliding flight, however, are albatrosses. These birds can travel for days without flapping their long, straight wings, using sea breezes and the currents of air that rise up from each wave to generate lift. Albatrosses, and other large seabirds, such as pelicans, make use of an aerodynamic phenomenon known as ground effect in order to achieve their lengthy glides. By flying low over a flat surface, such as the sea, the birds gain significant reductions in the energy they need to expend on flight, allowing them to glide effortlessly for long periods.

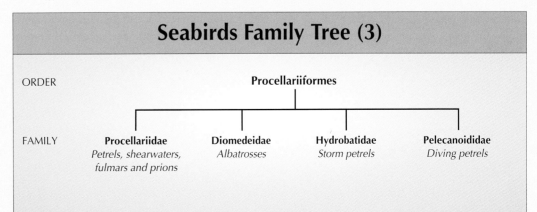

Seabirds Family Tree (3)

ORDER		Procellariiformes		
FAMILY	**Procellariidae** *Petrels, shearwaters, fulmars and prions*	**Diomedeidae** *Albatrosses*	**Hydrobatidae** *Storm petrels*	**Pelecanoididae** *Diving petrels*

Breeding behavior

Most seabirds conform to a rather generalized pattern of coloration, being dark on the dorsal (top) surface and white on the ventral (front). This probably provides camouflage, making them dark against the water when seen from above, and light against the sky when seen from below. However, many seabirds, such as the frigate birds and cormorants, have bright colors on the head, bill, or throat that play an important role in courtship. Some, such as puffins, acquire their colors only during the breeding season.

Courtship in seabirds can often be elaborate. Albatrosses undergo a highly ritualized dance, involving waggling movements of the head, wing stretching and bill clacking, accompanied by loud groans. These actions serve to strengthen the bond between the pair. Male roseate terns, *Sterna dougallii*, present their mates with a gift of a fish before mating begins.

Many of these birds are rather ungainly on land, and they and their eggs are vulnerable to predation. As a result, most species nest on offshore islands or precipitous cliffs, where they are relatively safe from foxes, rats and other predators. Suitable nesting areas are uncommon, and consequently most seabirds nest in large colonies, often sharing their nesting areas with other species. Nest space is usually very limited, and the birds can be highly territorial. Many, such as tropic birds, make no nest, laying their eggs on the bare rock or soil. Others, such as the gannets, construct a nest of mud and seaweed, adding to it each year, while pelicans build rudimentary nests from sticks. Puffins, prions and petrels burrow a hole into soft soil, where the eggs are laid.

Albatrosses lay only one egg, and some breed only every other year. However, they can live for more than 40 years. Most seabirds lay one or two eggs. Predation pressure on seabirds is generally low, and they can afford to rear small numbers of young each season. Also, almost all seabird chicks are fed on regurgitated fish and squid from the crop of an adult; it may be that not enough food can be supplied in this manner to provide for larger numbers of chicks. The largest chick sometimes dominates in the nest, and in some seabirds, such as boobies, it may actually kill its nest mates.

Threatened but surviving

In the past, seabirds were relatively free from predation pressure, both at sea and at the nest site. However, with the advent of commercial fishing and whaling, many island refuges became viewed as convenient stop-off points where the birds could be caught and eaten by sailors.

Direct hunting, as well as the release of animals such as rats, led to a dramatic decline in the numbers of many seabirds and the extinction of some, notably the great auk, *Pinguinus impennis*, from the North Atlantic. The population of the cahow, *Pterodroma cahow*, a shearwater from the western Atlantic, was severely depleted by settlers on the island of Bermuda where it nested, and the bird was believed to be extinct as far back as 1621. However, it was rediscovered in 1916, and a small colony was found nesting in 1951, although numbers remain critically low. Overfishing may result in starvation for the birds or their chicks, and in recent years albatross numbers have crashed dramatically. Many are trapped by long-lines set for tuna.

These northern gannets, Morus bassana, *were photographed near Bass Rock, off the east coast of Scotland, one of the largest breeding colonies of this species in the world.*

Other human activities have brought several species of seabirds to the brink of extinction. Seabird droppings, or guano, are rich in phosphates and make excellent fertilizer. However, the disturbance associated with guano mining has caused the decline of many species. The decline in seabird species has a number of other contributory factors. Short-tailed albatrosses, *Diomedea albatrus*, were hunted to the brink of extinction because of the trade in their feathers, which were used in hat making. Other birds, including many species of pelicans and cormorants, have been persecuted because they compete with fishing fleets for increasingly scarce fish stocks.

A few of the more adaptable seabirds have actually benefited greatly from human activities. The northern fulmar, *Fulmarus glacialis*, feeds extensively on offal from fishing vessels and has increased its range greatly. Many species of gulls can now be found miles away from the sea, feeding at garbage dumps and on other sources of urban waste.

For particular species see:
- ALBATROSS • BLACK-HEADED GULL • BOOBY
- CORMORANT • DARTER • DIVING PETREL
- FRIGATE BIRD • FULMAR • GANNET • GUILLEMOT
- HERRING GULL • JAEGER • KITTIWAKE • LITTLE AUK
- PELICAN • PRION • PUFFIN • RAZORBILL
- SHEARWATER • SHEATHBILL • SKIMMER
- SNOW PETREL • STORM PETREL • TERN • TROPIC BIRD

SEA BUTTERFLY

SEA BUTTERFLIES ARE transparent sea snails that are sometimes found swimming in large numbers near the surface of the sea. The 100 species are known collectively as pteropods because they were once classified together in a single order, the Pteropoda. In fact, there are two kinds of sea butterflies. Although both swim by means of winglike extensions of the foot, they are so dissimilar in other ways that they have clearly evolved separately from creeping ancestors. Members of one order, the Thecosomata, retain their shells as adults and feed on microscopic waterborne organisms by means of cilia (beating organs). Most members of the other order, the Gymnosomata, have a small, fragile shell only during the larval stages. This shell is shed during development, and the bodies of the adults are naked (the name of the order, from the Greek, means naked-bodied). Adult gymnosomes bear various kinds of armaments for the capture of large prey.

The shell of a thecosome may be spiral- or cone-shaped, and may be transparent and chitinous instead of calcareous. There may or may not be an operculum, the bony plate that covers the opening of the shell. The naked bodies of the gymnosomes are made up of two regions. The front region bears two weak eyes and various kinds of tentacles, as well as the winglike structures and a reduced foot. A plump, elongated trunk extends behind the front region. Within both orders there is much variety in appearance.

Grazers and predators

Sea butterflies are generally more abundant in tropical waters than in temperate seas. There are two notable exceptions, one species from each of the two groups, which are common in the waters of the eastern North Atlantic. One of these, the thecosome *Limacina retroversa*, is like a little winged snail with a left-handed spiral shell. Its maximum diameter is ⅛ inch (0.3 cm) but it is usually much smaller. Vast swarms occur in the North Sea and off the west coast of Scotland, particularly where oceanic and coastal waters mix. They are important as food for young herrings and other fish. *L. retroversa* is often preyed upon by the other exceptional cold-water sea butterfly, a common naked pteropod named *Clione limacina*. This predatory, winged gymnosome is delicately tinged with salmon pink on the wings and tentacles.

Limacina swims in a spiral course toward the surface, with the opening of its shell pointing upward and its wings beating in unison, like a butterfly. It then sinks a little way, with its wings

The sea butterfly **Clione** *is a sea snail that has lost its shell and taken to a pelagic (free-swimming, oceanic) life. The winglike projections, or swimming lobes, that propel the animal can be clearly seen.*

The naked beauty of the sea butterfly Clione limacina. *Although its sight is poor, this mollusk uses an array of specialized organs to prey on other pelagic mollusks.*

held still, before once again spiraling upward. As it moves in this way, prey items, mainly diatoms and dinoflagellates, are caught on the cilia covering parts of its body and are carried to its mouth. The ciliated food-catching surfaces are on the wings or foot in many species, but in *Limacina* especially they are found in the mantle cavity within the shelter of the shell. Thecosomes are usually found near the surface after twilight and start to descend again at midnight.

Ferocious butterflies

Gymnosomes are usually active near the surface by day, but tend to descend a short way at night. They feed on small animals in the plankton. Their wings beat more rapidly than those of the grazing thecosomes, and their swimming is faster and more maneuverable. The gymnosome *Clione* is armed with three pairs of conical projections that can be protruded from the mouth. It also has a pair of muscular pouches opening at each side of the gullet, lined with about 15 hooks that can be pushed out to grasp the prey. Most gymnosomes have these hook sacs, but the members of the Pneumodermatidae family add to this armor two, three or more sucker-bearing tentacles near the front of the body.

Although both kinds of sea butterflies are hermaphrodites (produce both eggs and sperm), cross-fertilization is necessary. The eggs, only 0.1 millimeter across, are laid in floating strips of transparent jelly, each about 2 millimeters long. *Hydromyles*, a naked pteropod of the Indian Ocean and South Pacific, broods its young. They lie first in an internal brood pouch and then in the general body cavity of the parent, finally being released as the body of the parent degenerates and ruptures. The more usual method is for the eggs to be laid in the sea, from which hatch typical molluskan veliger larvae. Each has a thin shell from which protrude its ciliated, membranous body parts. In the gymnosomes the shell is discarded early, and the veliger stage is followed by a larva with rings of cilia.

SEA BUTTERFLIES

PHYLUM	**Mollusca**
CLASS	**Gastropoda**
ORDER	**Gymnosomata and Thecosomata**
FAMILY	**Clionidae; others**
GENUS	***Clione, Limacina*; others**
SPECIES	**100, including *Clione limacina* (described below)**

ALTERNATIVE NAME
Pteropod (general term for all species)

LENGTH
Cold-water subspecies: up to 3 in. (8 cm); warm-water subspecies: up to ½ in. (1.2 cm)

DISTINCTIVE FEATURES
Sea snail with translucent, naked body; foot has winglike extensions; 2 tentacles at front

DIET
Large plankton, including other sea butterflies

BREEDING
Hermaphroditic (produce both eggs and sperm); fertilized eggs shed into water in jellylike masses; several larval stages

LIFE SPAN
Probably less than 1 year

HABITAT
Open waters, from surface to 490 ft. (150 m)

DISTRIBUTION
Colder waters of North Atlantic and northeast Pacific Oceans

STATUS
Abundant

Ocean bed builders

The shelled pteropods are so numerous in some places that the seabed becomes covered with a thick accumulation of their shells, forming what is known as pteropod ooze. This is most common near coral islands at depths of 2,400–9,000 feet (730–2,740 m). The ooze does not occur in deeper water because the shells there take so long in sinking that they are dissolved long before they reach the seabed.

Fishers also know the sea butterflies, but indirectly. Where *Limacina* is abundant, herring may eat them to the exclusion of their usual food. Their internal organs may take up so much of the dark sepia stain produced by these mollusks that they get what fishers call black gut.

SEA CUCUMBER

SEA CUCUMBERS ARE related to starfish and sea urchins. Many have a cucumberlike shape, but a few of the 1,000 species have a more unusual appearance. Their common name derives from their use in Asia for soups.

As with starfish, the sea cucumber's body parts are arranged in fives or multiples of five. At the front end surrounding the mouth is a crown of tentacles, usually numbering 10 or 20 although there may be as many as 30. The long, almost cylindrical body of a typical sea cucumber may be slightly flattened underneath. It has five longitudinal ambulacral areas (regions containing the principal nerves, blood and water-carrying vessels), each containing two or more rows of tube feet. Often the tube feet are scattered over three grooves on the underside and are used to creep over the seabed. The tube feet on the upper side, running along the back, are small, fewer in number and often degenerate. A few species have no tube feet. Most sea cucumbers are several inches to 1 foot (30 cm) long, but a few are several feet long. They are mainly uniformly colored, occasionally paler on the underside, sometimes with darker markings. The colors are usually gray, brown, black or a shade of purple, rarely red or orange, but some small burrowing forms may be pink or violet.

There are five orders that differ mainly in the shape of their tentacles and tube feet. Sea cucumbers are found in all seas, usually at depths of less than 600 feet (180 m), but a few live at very great depths, down to 33,000 feet (9,900 m).

Shapes for all situations

Some sea cucumbers creep about, mainly over sandy seabeds, whereas others burrow in sand. Some burrowing sea cucumbers have both ends of the body upturned, so that the animal is permanently U-shaped. This leaves the mouth and tentacles at the surface for feeding and the hind end at the surface for ejecting waste. Some burrowing forms have a long tail-like part that curves up to the surface and gets rid of waste.

The sea cucumber *Rhopalodina* has a bulbous body with the mouth and anus at the end of a long stalk that may be three or four times the length of the body. Some sea cucumbers are flattened, with spinelike processes around the margins of the body. *Psolus diomediae* resembles a large slug. Nearly all sea cucumbers live on the

seabed, but one species, *Galatheathuria aspera*, is flattened and swims freely in the sea. Another, *Pelagothuria natatrix*, is octopuslike, with 12 arms. The deep-sea *Scotoplanes globosa* has a rounded, oval body and two pairs of bent appendages, one pair pointing forward, the other backward.

Most sea cucumbers move over the seabed using their tube feet. Some have no tube feet and move by muscular contraction and expansion of the body, like an earthworm. To aid movement, they may grip the seabed with their tentacles and use spicules (microscopic rods and plates of calcium that make the skin rough) in their skin. The rods may be branched or may form anchors; the plates may be perforated or form wheels, baskets or tables. The skin of a sea cucumber 3 inches (7.5 cm) long may contain 20 million spicules.

Feeding techniques

Sea cucumbers feed on microscopic organisms or on detritus. One common European species, *Cucumaria saxicola*, lives in burrows made by other animals in rocks, and stretches out its tentacles into the water to feed. Small floating organisms get stuck on the slimy tentacles, which are then bent back to the mouth. Other sea cucumbers, such as *Holothuria*, use their short tentacles to scoop sand and mud into the mouth. Burrowing sea cucumbers swallow sand or mud as they plow through it, digesting any edible matter and ejecting the rest. Studies have shown there may be 2,000 sea cucumbers per acre of coral reef, passing 60 tons (54 tonnes) of sand through their bodies each year.

Many sea cucumbers move huge amounts of sand and mud as they feed, making them significant members of the coral reef community. The species pictured is in the genus Holothuria.

Free-swimming larvae

Most sea cucumbers shed their eggs and sperms into the water. The fertilized eggs develop into free-swimming larvae called auriculariae, which are semitransparent and irregularly shaped, as if they have been much folded. They have a number of lobes or arms with a simple gut at the center. Around the margins of the body and arms is a continuous band of cilia, which drive the larvae through the water. Later the arms are withdrawn, giving the body a barrel shape, and the cilia form bands around it. Then the bilaterally symmetrical larvae, instead of swimming straight forward, spin like tops. After a while the larvae sink to the seabed, each growing five tentacles to become small sea cucumbers.

Some female sea cucumbers carry their eggs in separate pockets along the back; others have larger pouches with several eggs in each. In a few species the females incubate their eggs inside the body, and one species carries the eggs in pockets in the stomach.

Defense and symbiosis

The main predators of sea cucumbers are probably starfish, although they are also eaten by bottom-feeding fish. However, they have tough, slippery skins to protect themselves. Some give out a poison when they become alarmed, while others squirt out sticky white threads through the anus. One species, *Holothuria forskali*, is called the cotton spinner because of this habit. The threads entangle the attacking animal and, once used, they cannot be drawn in again. The sea cucumber merely grows a new set. Some forms throw out their internal organs at an attacker and grow a new set afterward.

Larger sea cucumbers are caught in East Asia and off the coasts of northwestern Australia. They are eaten raw, cooked as soup, or boiled and dried to be eaten later as *trepang* or *bêche de mer*.

A host to other animals

Some animals, including pearlfish and several types of bivalve mollusks, use sea cucumbers as homes. One bivalve, *Devonia semperi*, found in the seas around Japan, clings to the surface of sea cucumbers using a suction-cup-shaped foot. Also around Japan, there is a sea snail, *Parenteroxenos dogieli*, that has no shell or internal organs except those for reproduction. It feeds directly from its host, *Cucumaria japonica*. It is wormlike, much coiled and, when stretched out, measures 4 feet (1.2 m), dwarfing its 4-inch (10-cm) host.

Sea cucumbers are bilaterally symmetrical. Some species, such as Cucumaria miniata (below), spit out sticky threads when disturbed, in order to discourage predators.

SEA CUCUMBERS

PHYLUM	**Echinodermata**
CLASS	**Holothuroidea**
SPECIES	**About 1,000, including *Pelagothuria natatrix*; *Galatheathuria aspera*; *Psolus diomediae*; *Cucumaria saxicola*; *C. miniata*; *C. japonica* and *Holothuria forskali***

LENGTH
Variable, from a few inches to several feet

DISTINCTIVE FEATURES
Most species resemble very fat, elongated cucumbers; some species wormlike; mouth surrounded by up to 30 tentacles, in multiples of 5; several rows of small tube feet on underside and upper side

DIET
Omnivorous, including detritus; burrowing sea cucumbers ingest sand with food

BREEDING
Sexes separate in most species; some species hermaphroditic (producing sperm and eggs); fertilized eggs hatch into free-swimming larvae; metamorphose after several weeks

LIFE SPAN
Not known

HABITAT
On sand, mud, detritus, coral reefs and rocks; from shallows to extreme depths

DISTRIBUTION
Seas and oceans worldwide

STATUS
Many species common, but heavily fished

SEA FIR

EA FIRS, OR HYDROIDS, resemble tiny, delicate seaweeds. In fact, they are colonies of minute polyps related to sea anemones and corals. There are more than 2,000 species. The smallest is about 3 millimeters tall, though a few deep-sea species grow to 6 feet (1.8 m) tall; one deep-sea species in the genus *Branchiocerianthus*, living at a depth of 15,000 feet (4,500 m), consists of one polyp 8 feet (2.4 m) tall.

One sea fir genus, *Obelia*, is found worldwide and illustrates the general pattern of branching in sea firs. A tuft of this sea fir is made up of a horizontal network of branching roots, from which vertical stems arise. Each has a zigzag pattern, with a horny cup at each angle of the zigzag. The whole of the roots and stems consists of horny tubes, with a tubular strand of living tissue running throughout but expanding in each horny cup to form a polyp. There are other sea fir species in which the vertical stems do not branch, each being a horny tube with a single polyp at the top, and others that have no horny coverings. These last have a horizontal network of tubular roots from which vertical stems grow, each ending in a polyp, but without the protective horny sheath. They are called naked sea firs.

Sea firs are found in all seas, usually between tidemarks or in shallow seas, with few species living at depths of more than 330 feet (100 m). In the shallow seas and on the shore, sea firs grow on any solid support on the seabed, such as seaweeds, shells, pebbles, rocks, crabs or wrecks. Sea firs will also grow on a sandy bed, with their roots buried in the sand. A few species, such as those in the genus *Cordylophora*, live in brackish water in estuaries.

Sea fir commonwealth

Sea firs are related to the freshwater hydra. They have stinging cells for catching small animals, and in most other ways their polyps live like hydra. But whereas the latter consist of a single polyp that sometimes has young hydra, formed from buds, on the sides of its body, sea firs multiply by budding and the buds remain continuous, forming a branching colony. Otherwise hydra and sea firs have a very similar way of life except for their method of sexual reproduction. In both sea firs and corals, in which there are many small polyps all in communication via a system of tubes, a form of simple commonwealth exists. Food obtained by any one polyp helps to nourish all the others. Therefore, it is not necessary for all the polyps to get food. The colony flourishes as long as enough of its polyps catch food.

Making the zigzag

After a sea fir larva settles on the bottom, it soon loses its shape and becomes a tube that begins to grow over the surface of the rock or seaweed. It also begins to branch, and at the same time a stalked polyp grows vertically from one point on the tube. As soon as this polyp is formed, another buds out just below it, but this points in the opposite direction. When this polyp is fully formed, another buds out below it, in the other direction. In this way the stem zigzags upward. Meanwhile, the horizontal rootlike bases spread out and at many points along them other vertical stems arise and grow by budding.

Alternation of generations

The colony of polyps that forms a sea fir is produced asexually, by budding. Once it is well grown, another reproductive method takes over.

The featherlike sea fir Plumularia setacea grows in two forms. Settled on other hydroids, it reaches about ⁶/₁₀ inch (15 mm). On rocky substrates it may grow to 2¾ inches (7 cm) or more.

Sea firs do not always grow upward. The group of Sertularella speciosa *shown above are growing from the underside of an algae-covered crevice.*

SEA FIRS

PHYLUM	**Cnidaria**
CLASS	**Hydrozoa**
ORDER	**Hydroida**
FAMILY	**Many, including Tubularidae, Corynidae and Clauidae**
GENUS AND SPECIES	**More than 2,000 species, including *Tubularia larynx* (detailed below); *T. indivisa*; and *T. bellis***

LENGTH
Up to 2 in. (5 cm)

DISTINCTIVE FEATURES
Dense colonies of polyps, which resemble tiny, delicate seaweeds; loosely branched; often pink or rose red in color

DIET
Small organisms

BREEDING
Both sexual (takes place during free-living medusa stage) and asexual (takes place when organism is attached)

LIFE SPAN
Possibly several years or more

HABITAT
Rocks, stones, shells and seaweeds, often where water movement is stronger; from low shoreline to depth of up to 330 ft. (100 m)

DISTRIBUTION
Not known; possibly worldwide

STATUS
Common

In *Obelia*, for example, a horny, urn-shaped structure grows out at intervals on the zigzag vertical stems, with an opening at the top and a rod of living tissue through the middle. Buds begin to appear on this rod, and each grows into a very tiny jellyfish that finally breaks loose and swims out into the sea. On its undersurface are four groups of sex cells, male in some jellyfish, female in others. The sperms and ova are shed into the sea, where the ova are fertilized. They develop into swimming larvae that settle on the bottom, where they grow into stalked colonies of polyps.

This constitutes an alternation of generations. The sea fir is the asexual generation, which alternates with the jellyfish or sexual generation. Not all sea firs produce free-swimming jellyfish. In some the jellyfish remain attached to the parent stock, the eggs remain in the jellyfish and the young are liberated not as larvae but as young polyps with a ring of tentacles around the mouths. In other species the jellyfish or sexual generation is suppressed. In *Dynamena* (also known as *Sertularia*) the urn-shaped cups contain ova that develop straight into larvae after they have been fertilized.

Why sea firs?

Sea firs were first studied scientifically by John Ellis, an 18th-century English naturalist, in his *History of Corallines*, published in 1755, and again in his *History of Zoophytes* of 1785. Ellis used both the term zoophytes and the term corallines to describe the organisms, but later writers used zoophytes only, and late in the 19th century the name hydroids became universally adopted. The name zoophytes comes from the Greek language and means animal-plants; Ellis clearly seems to have regarded them as such. He gave the various species different names, including sea cypress, sea tamarisk, sea oak, sea fern, lily-flowering coralline and sea spleenwort. However, he called one species *Sertularia abietena*, or sea fir, because he thought it resembled a miniature fir tree, and the name gradually became accepted to describe other organisms of this kind.

SEA HARE

THE SEA HARE IS A MOLLUSK and bears similarities to both the sea slug and the sea snail. Its common name dates back to the Roman naturalist Pliny, who detected a resemblance between the mollusk and a crouching hare. A sea hare has a soft body that collapses when it is removed from water but is graceful when fully immersed. The foot is folded up on each side, hiding the small, thin transparent shell. The presence of a shell is the main difference between sea hares and sea slugs (nudibranchs). The body is long, and on the head are two pairs of tentacles, the upper pair of which bears some similarity to the ears of a hare.

In European waters sea hares are usually about 3 inches (7.5 cm) long and olive green, brown or reddish in color. Species remarkably like them are found in coastal waters in most temperate or tropical parts of the world. Along the Pacific Coast of North America, for example, there are sea hares, sometimes called sea rabbits, that have the same shape and are also olive green to brown. They grow much larger, however: 12 inches (30 cm) or longer with average weights of 7–8 pounds (3–3.5 kg), although some may weigh as much as 16 pounds (7 kg). On the Australian coasts there is a similar species, with individuals 12 inches or longer, olive green with small, dark brown or black circles or with black mottlings and streaks; another, rarer, sea hare is black. There are also other species in Australian waters that move more slowly than usual and have a rougher surface with many small warts. One of these is about the size of the European sea hare and is light green; the other is nearer the size of the Pacific American species and is dark green.

Double gizzards

Sea hares browse on seaweeds, the European species feeding preferentially on the green sea lettuce *Ulva*, as well as on *Enteromorpha* and the red *Plocamum*. They can feed on many other seaweeds, but grow much more slowly when they do so. The food source also affects the sea hares' color. The larger species elsewhere feed on larger, coarser seaweeds. Using the rasplike surface of the horny tongue or radula, the sea hares scrape off fragments of weed, which are swallowed and passed to the first of three stomachs. This receives the food and passes it on to the second stomach, which is lined with horny teeth and acts more like a gizzard than a stomach, tearing the fragments of seaweed into smaller shreds. The third stomach is similar to the second and further breaks up the food.

The two fleshy flaps that are folded upward over the shell are used to create currents of water that flow into the gill cavity, where there is a true gill. Sea hares breathe, therefore, by flapping their "wings," so that water is taken in. Sea hares can also use these wings to swim, although they rarely do so.

Microscopic eggs

In summer, sea hares come up onto the beaches to lay their eggs. In some places they invade the beaches in their thousands. Having done so, they return to the shallow waters below the low tidemark and soon die. Most species have a life span of a year, and the longest-lived sea hares do not survive more than 2 years.

There is no obvious reason why sea hares should come so far up the beach to lay their eggs on the seaweeds and rocks. It may be that the temperatures are higher there, or the water flowing over the beach is better aerated. Juvenile sea hares are often found feeding in weedy rock pools, which may make good nurseries.

Sea hares, such as this pair of the species Aplysia punctata, *migrate inshore to lay their eggs, which may amount to many thousands over a number of months.*

Each sea hare is hermaphroditic (possesses both male and female organs), but is not self-fertilizing. At a single mating one of a pair of sea hares acts as a female, the other as a male, although usually the two act both as male and female in a single pairing. The coupling may last for hours or even several days on end; seven or eight sea hares have been seen mating in a circle.

The eggs are laid in strings of yellowish jelly produced at the rate of 2⅖ inches (6 cm) per minute. Such a length of string contains about 230 capsules, or 41,000 eggs. Each egg is 0.05 millimeters in diameter. These figures are for *Tethys californica* and are given by biologist G. E. MacGinitie, who counted the eggs by taking parts of the egg string, then weighing them on a microscale and counting every egg. He weighed the whole mass and estimated the total number of eggs. He found that a sea hare weighing 5 pounds 12 ounces (2.6 kg) laid 478 million eggs in 4 months and 1 week. There were 27 separate layings, the longest measuring ⅓ mile (0.5 km).

When laying eggs the sea hare grasps the jelly string in a fold of its upper lip and, as the string leaves the body, covers it with a sticky secretion from the lip. Then, moving its head from left to right, it sticks the string down at irregular intervals onto that already laid. At the end the total egg mass looks like a compact tangle of yarn fastened to seaweeds. Not all sea hares do this. Some lay their eggs on a rock surface in a close but regular zigzag. However, the method of laying is similar in all species. The free-swimming larvae hatching from the spawn in 12 days form part of the plankton but soon settle on the bottom as young sea hares.

Low survival rate

In the wild, only one or two of the many millions of offspring from each sea hare reach adulthood. The larvae are eaten by plankton-feeders, and the young fall prey to bottom feeders. Those that do reach the adult state are relatively safe, since they are distasteful to predators, of which there are virtually none.

Feared by the ancients

In ancient times Mediterranean populations believed that the sea hare had magical powers. This may have been partly because of the mollusk's shape. Alternatively, perhaps superstitions relating to the hare on land were transferred to the sea hare and elaborated.

This suspicion of the sea hare would probably have been enhanced by the unpleasant smell that it exudes and by the purple fluid it releases. Scientists do not know whether this fluid acts as a screen against predators, like the ink of the cuttlefish, or is a deterrent in some other way.

The California sea hare, Aplysia californica, *ejects purple ink as a form of self-defense.*

SEA HARE	
PHYLUM	**Mollusca**
CLASS	**Gastropoda**
ORDER	**Anaspidea**
FAMILY	**Aplysiidae**
GENUS AND SPECIES	***Aplysia punctata; others***

LENGTH
Up to 12 in. (30 cm)

DISTINCTIVE FEATURES
Internal shell, fragile, transparent amber; olive-green, brown or blackish body, often blotchy; enlarged lobes on each side of body; four distinct hornlike appendages on head

DIET
Seaweeds, especially green species such as *Ulva* and *Enteromorpha*

BREEDING
Hermaphroditic; eggs laid in spring hatch into free-swimming veliger larvae; these spend some time as plankton, then develop into adult form

LIFE SPAN
Up to 2 years

HABITAT
Variety of rocks; weedy and stony bottoms; probably most common in upper 66 ft. (20 m)

STATUS
Common

SEA HORSE

THE SEA HORSE IS AN unusual fish that looks almost like the knight of a chess set. It hangs suspended in the water, with its tail wrapped around seaweed or eelgrass. Another peculiar feature is that each of its eyes is on a turret that can move independently. Although many other fish can also move their eyes independently, this ability is more pronounced in sea horses. A final oddity is that the male carries the fertilized eggs, and later the baby sea horses, in a pouch.

A sea horse has a large head positioned, unusually for a fish, at a right angle to the body. It has a tubular snout, a mobile neck, a rotund body and a long, slender tail. Size ranges from the pygmy sea horse, *Hippocampus bargibanti*, and the tiny western Atlantic dwarf sea horse, *H. zosterae*, both of which are just ¾ inch (2 cm) long, up to the large White's sea horse, *H. whitei*, which measures 13¾ inches (35 cm) in total length. The neck, body and tail are marked with circular and longitudinal ridges, on which there are bony bumps, so the fish looks almost like a wood carving. There is a pair of small pectoral fins and a single small dorsal fin. There are often fleshy strands, which may serve as camouflage.

There are 32 species of sea horses, distributed worldwide. The majority of species live in Southeast Asia and Australasia. The others live off the Atlantic coasts of Europe, Africa and North America, with two species on the Pacific Coast of America.

Swimming upright

Sea horses live in shallow inshore waters among seaweeds or in beds of eelgrass in estuaries. They swim in a vertical position, propelling themselves by rapid waves of the dorsal fin. When they swim at full speed, this fin may oscillate at a rate of 35 times a second, making it look a bit like a revolving propeller. The pectoral fins oscillate at the same rate, and the head is used for steering, the fish turning its head in the direction it wants to go. They are able to rise or sink in the water by altering the volume of air in the swim bladder. If the fins are damaged, they can be regenerated relatively quickly.

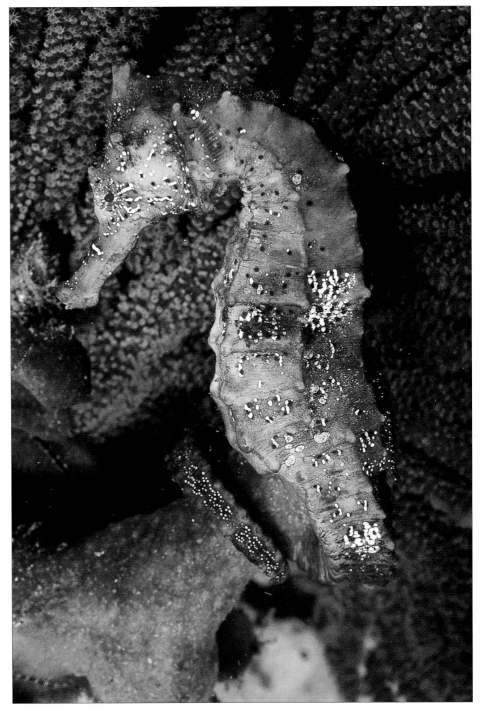

Pipette mouth

Sea horses eat their prey in an unusual manner. The long snout acts like a pipette, and the food is drawn in rapidly by a slight inflation of the sea horse's body. The fish approach their tiny prey in a very leisurely way, peer at it for a couple of seconds, and then, having placed their snout in a convenient position, suddenly engulf the meal. Prey is mainly tiny crustaceans such as copepods, but baby fish are also eaten.

Sea horses such as this Pacific or yellow sea horse, Hippocampus ingens, are among the weakest of swimmers. Even in slight currents they anchor themselves to gorgonian corals with their prehensile tails.

The curious body form of sea horses can serve both as armor and as camouflage, as in the case of the pygmy sea horse (above), one of the smallest species.

LONGSNOUT SEA HORSE

CLASS	**Osteichthyes**
ORDER	**Syngnathiformes**
FAMILY	**Syngnathidae**
GENUS AND SPECIES	*Hippocampus reidi*

ALTERNATIVE NAME
Longnose sea horse

LENGTH
Up to 6 in. (15 cm)

DISTINCTIVE FEATURES
Yellow to reddish-brown body; solid color, spotted or speckled

DIET
Zooplankton (animal plankton)

BREEDING
Breeding season: 8 months in the laboratory; number of eggs: 100 to 200; hatching period: 10–60 days, incubated in the pouch of the male

LIFE SPAN
A few years

HABITAT
Surface of gorgonian corals, *Zostera* eelgrass or floating *Sargassum* seaweed, attached by the tail, or free swimming in shallow water to 50 ft. (15 m)

DISTRIBUTION
Coastal waters of western Atlantic, from Nova Scotia south to São Paulo, Brazil

STATUS
Vulnerable

☐ Pacific sea horse   ■ Longsnout sea horse

Father carries the young

Breeding starts when a male finds a female and begins courtship. Depending on the species, either he swims in front of her without actually touching her, or the two entwine tails. He seems to be bowing to her, but this is actually a pumping action to drive the water out of the pouch on his belly. The female then inserts her long ovipositor into the opening of the pouch to lay her eggs. When laying is finished, the mouth of the pouch closes to a minute pore and stays like this until the baby sea horses are ready to be born. The young sea horses are ⅓–½ inch (0.8–1.3 cm) long at birth and are perfect miniatures of their parents. The first thing baby sea horses do is to swim to the surface and gulp air to fill their swim bladders. They then feed ravenously on extremely small crustaceans, such as newly hatched brine shrimps, and grow rapidly.

Placental fishes

The inside of the pouch changes just before and during courtship. The walls thicken and become spongy, and they are enriched with an abundant supply of blood vessels. As the female lays her eggs, the male fertilizes them and they become embedded in these spongy walls. The network of blood vessels in the wall of the pouch probably acts like a placenta, passing oxygen to the eggs and taking up carbon dioxide from them. Also, food probably passes from the paternal blood into the eggs, just as it does from the mother's blood in the mammalian placenta.

Male labor

We are accustomed to the idea that the actual bearing of offspring is always done by the female. In sea horses it is the reverse. As each batch of eggs is laid in his pouch, the male sea horse goes through violent muscular spasms. These work the eggs to the bottom of the pouch to make room for more. It seems also that there is a physiological reaction as the eggs sink into the spongy tissue, and he shows signs of exhaustion. When the young have hatched and are ready to leave the pouch, the mouth of the pouch opens wide. The male alternately bends and straightens his body in convulsive jerks, and finally a baby sea horse is shot out through the mouth of the pouch. After each birth the male rests. In aquaria the males often die after delivering their brood, but this does not seem to happen in the wild.

Hippocampus hippo-campus, *which lives in the Mediterranean Sea, can be recognized by its short snout. This one has chosen a peacock worm as an anchorage point.*

SEA LILY

The sea lily Newmaster rubiginosa *of the Caribbean Sea has branched arms, the outer parts of which have a membranous cellular tissue that produces the sex cells.*

OF ALL ANIMAL SURVIVORS from the past, probably none has a greater claim than the sea lily to being called a living fossil. Although related to starfish, sea urchins and sea cucumbers, sea lilies are plantlike in shape. In a typical sea lily the body is at the top of a long, slender stalk. At the base of the stalk in some species the rooting processes that anchor the sea lily to the bottom are slender and branching and look like the roots of a land plant.

The body itself is small and bears five arms, each of which has a row of pinnules (secondary branches) so that the sea lily looks more like a miniature palm tree than an animal. Most commonly the arms are doubled at the base, producing what appear to be 10 arms. The greatest number of arms in any sea lily is 56. In some species the arms have many branches, each having its rows of pinnules. In some the stalk is plain, while in others it carries throughout its length many tentacle-like processes, called cirri, the precise function of which is not known. The colors of sea lilies are mainly yellow, pink and red. The smallest sea lily living today is only 1½ inches (3.75 cm) high and the largest is almost 39 inches (1 m) high, although some fossil sea lilies were as much as 70 feet (21 m) high.

The closest relatives of sea lilies are the featherstars, found mainly in shallow seas. However, featherstars have no stalk and are free moving, attaching themselves temporarily to the bottom using clawed processes. Sea lilies live in most oceans but rarely in waters less than 600 feet (183 m) deep. They are most numerous at depths of about 3,600 feet (1,080 m); some species are found as deep as 27,000 feet (8,100 m).

Bodies mainly chalk

There are few fossils in rocks earlier than the beginning of the Cambrian period, which began around 600 million years ago, but there are many remains of sea lilies from the start of that period. From then on, down through the geological ages, sea lilies continued to be numerous, so much so that in some places limestone rocks are made up largely of their skeletons. Usually these skeletons are broken, but many sea lily fossils have been collected in an almost perfect state.

Today there are about 80 known species of sea lilies worldwide, from at least 35 genera. Like their ancestors, they are almost 90 percent chalk, which explains their alternative name of stone lilies. The stalk is made of calcium carbonate joints that look like simple vertebrae, although sea lilies are invertebrates. The resemblance to vertebrates is only superficial. The small body is strengthened with chalky plates, and the arms and their branches have a skeleton of smaller, vertebralike joints running through their centers. The living tissue makes up only a small percentage of the total bulk of the animal, and the internal organs are similar to those of a starfish.

Like the starfish, the sea lily also has nerves and muscles, as described by American oceanographer Louis Agassiz (1807–1873), who was probably the first person to witness a living sea lily brought up from the depths: "When disturbed the pinnules of the arms first contract, the arms straighten themselves out, and the whole gradually and slowly closes up. It was a very impressive sight for me to watch the movement of the creature, for it not only told of its own ways, but at the same time afforded a glimpse into the countless ages of the past, when these crinoids… so rarely seen today, formed a prominent feature of the animal kingdom."

Sensitive to change

One reason scientists know so little about the way of life of sea lilies is that they are deep-sea animals, so that scientists are not able to observe them directly. On the rare occasions that living specimens are brought in, they last very little

SEA LILIES

PHYLUM	**Echinodermata**
CLASS	**Crinoidea**
ORDER	**Millericrinida, Cyrtocrinida; others**
FAMILY	**Many, including Bathyarinidae**
GENUS	**At least 35 genera**
SPECIES	**About 80, including *Bathycrinus carpenterii* (described below)**

ALTERNATIVE NAME
Stone lily (name applied to all species)

LENGTH
Up to 10 in. (25 cm)

DISTINCTIVE FEATURES
Resembles miniature palm tree with 5 pairs of arms, each with further small branches

DIET
Small particles of decaying material

BREEDING
Sexes separate, but no distinct sex organs; gametes (sex cells) develop along arms; arm walls rupture to release eggs or sperm into water for fertilization

LIFE SPAN
Not known

HABITAT
Deep-sea muds and oozes

DISTRIBUTION
Northern Atlantic Ocean

STATUS
Locally common

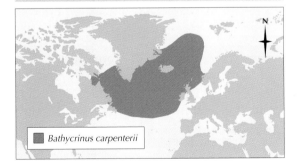

Bathycrinus carpenterii

These sea lilies, Analcidometra caribbea, are attached to a sea fan or gorgonian coral of the genus Elissela.

naturalists have established a working knowledge of their manner of feeding, although more by deduction than by direct observation. The arms are grooved and lined with cilia. The sea lily may spread its arms and pinnules to form a netlike structure that catches small living organisms or particles from the decaying bodies of small dead animals slowly raining down from the surface waters. These particles, trapped in the grooves, are passed on to the mouth by the cilia.

Discovery from the depths

For a long time scientists assumed that the sea lilies were extinct. Even when the first living specimens were found they were thought to be rare until oceangoing research cruisers began to haul up their dredges filled with sea lilies. The USS *Albatross*, for example, made a haul on one occasion of 3 tons (2.7 tonnes) of sea lilies in one sweep of the dredge. A later United States expedition organized specially to search for sea lilies in the early 20th century found forests of them in the ocean depths of the Pacific. The first sea lily was found in 1755, in deep water off Martinique in the Caribbean. It was sent to France and in 1761 was exhibited before the French Academy of Science as the *Palma marina*, or sea palm. The second specimen was brought up from deep water off Barbados a few years later. This was exhibited before the Royal Society in London. One hundred years later another sea lily was dredged up from deep water off the Lofoten Islands off Norway. It inspired as much scientific attention as the discovery of the coelacanth (discussed elsewhere) did 70 years later.

time in aquaria unless conditions are kept absolutely right for them. The slightest adverse change in circumstances causes them to break up, and it is probably because they are so sensitive to the slightest change or disturbance in the water that they live at depths where there is no turbulence. Although much of the natural history of sea lilies remains a matter of scientific debate,

SEA LION

THE SEA LION THAT traditionally featured in circus performances is the California sea lion, *Zalophus californianus*, the smallest of the five species, all of which resemble the fur seals (discussed elsewhere). Both sea lions and fur seals have small external ears and both turn the hind flippers forward to move by bounding on their flippers. Sea lions, however, have broader muzzles than fur seals. Male California sea lions measure 7 feet (2.1 m) and females about 6 feet (1.8 m); the males lack the lionlike mane characteristic of other sea lions. When the fur is wet it appears black, but it dries to a chocolate brown. California sea lions are found on the coasts of California and northern Mexico, on offshore islands, on the Galapagos Islands and on islands off the Japanese coast.

The other northern sea lion is Steller's sea lion, *Eumetopias jubatus*, which lives in the northern Pacific from Japan around to California. Adult males measure about 11 feet (3.3 m) and weigh around 1 ton (1,015 kg). Females are very much smaller, measuring about 7½ feet (2.3 m) and weighing only 600 pounds (270 kg). Adult males develop very thick necks and have shaggy manes. The southern sea lion, *Otaria flavescens*, is similar to Steller's sea lion but is smaller, except for the Australian sea lion, *Neophoca cinerea*. The Australian sea lion is found on the coasts of southwestern Australia. Hooker's sea lion, *Phocarctos hookeri*, is restricted to Auckland, Campbell and Snares Islands, all to the south of New Zealand. The southern sea lion is found on the coasts of South America from northern Peru, around Cape Horn to southern Brazil and the Falkland Islands.

Outside the breeding season sea lions live in large mixed herds on rocky shores, but some migrate. California sea lions and Steller's sea lions often move northward in winter, some of the latter reaching the Bering Strait, but they return south when the sea freezes.

Cartilaginous ribs

Except when they are defending their pups or territories, sea lions are usually quite tame and it is possible to walk close to them. They are generally wary rather than aggressive, and sometimes a whole herd will panic and rush into the sea. On reasonably smooth ground they can outpace a

Depending on their geographical location, California sea lions breed between June and September each year, usually producing a single pup.

SEA LIONS

CLASS **Mammalia**

ORDER **Pinnipedia**

FAMILY **Otariidae**

GENUS AND SPECIES **California sea lion,** *Zalophus californianus;* **Steller's sea lion,** *Eumetopias jubatus;* **southern sea lion,** *Otaria flavescens;* **Australian sea lion,** *Neophoca cinerea;* **Hooker's sea lion,** *Phocarctos hookeri*

WEIGHT
Up to 2,500 lb. (1,120 kg)

LENGTH
Head and body: 5–11 ft. (1.5–3.3 m)

DISTINCTIVE FEATURES
Streamlined body; blunt snout; small ears; powerful flippers; short, dark brown coat

DIET
Fish, squid and crustaceans

BREEDING
California sea lion. Age at first breeding: 6–9 years; breeding season: June–September; number of young: usually 1; gestation period: 11 months, including 3-month delayed implantation; breeding interval: 1 year.

LIFE SPAN
Up to about 35 years

HABITAT
Coastal waters; rest and breed on islands and sandy or rocky shores

DISTRIBUTION
West Coast of U.S.; Alaska; northwestern Mexico; Galapagos Islands; South America; Japanese islands; southwestern Australia

STATUS
Hooker's sea lion: vulnerable. Steller's sea lion: endangered. Other species: locally common to uncommon.

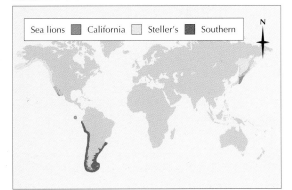

Sea lions ■ California □ Steller's ■ Southern

Sleek and streamlined, and with large front and hind flippers, Australian sea lions are perfectly adapted for their underwater activities.

human, although they are less maneuverable. In the Falkland Islands sea lions trample paths through the tall, dense tussock grass and, when disturbed, they rush down these paths to safety in the sea. Sea lions display remarkable agility in leaping over rocks and down steep slopes, and may even jump over a low cliff. The shock of the fall is absorbed by the front flippers, the blubber and the soft, cartilaginous ribs, a real advantage when swimming around rocks in heavy seas.

Eyesight and echolocation

Although their eyesight is bad, sea lions are expert fishers. In murky water they probably use their long whiskers to detect their prey and other objects. Recently it has been found that sea lions, and probably other seals, use echolocation, or sonar, as well. Steller's sea lion, which scientists have studied in detail, eats a wide variety of food, including squid, herring, pollack, halibut, sculpin and salmon. Wherever sea lions live near a commercial fishery they are blamed for damage both to fish and to equipment. In the Alaska

The number of Steller's sea lions in the Alaska region fell by 85 percent in 35 years, from 230,000 in 1965 to 34,000 in 2000. The decline was caused by overfishing of the sea lions' main prey by the local fishery.

region, for example, there is a battle between environmentalists, who are trying to save Steller's sea lions, and the U.S. fishing industry. A rise in the catch numbers of pollack has matched a regional decline in the number of sea lions. An injunction has been granted to prevent fishers trawling in areas off the Alaskan coast.

California sea lions, also unpopular with fishers, appear to prefer squid to salmon. Around the Falkland Islands, southern sea lions eat fish although crustaceans and squid are the main food.

Breeding habits

At the start of the breeding season, the mixed herds split up as each mature bull attempts to stake out a territory. As with elephant seals and fur seals, only the old bulls can form a territory; the younger bulls are driven off the beach and spend their time just offshore or on common ground where the sea lions mix.

Each bull gathers a harem of 10 to 20 cows in his territory. The cows come ashore 2–3 weeks after the bulls, when the territory boundaries have been decided and most of the serious fighting has finished. Within a day or two the

single pups are born. A few days later the cows mate again with the bull in whose territory they have pupped and then return to the sea to feed, returning every so often to suckle their pups. The pups swim at an early age and are suckled for up to a year. Many pups die in storms, although their mothers carry them by their scruffs away from the waves. Depending on the species, females are sexually mature at about 6 years and males at 9 years, when they are large enough to hold a territory.

Intelligent performers

Many seals are naturally playful, and sea lions are especially so. They have even been observed chasing their own streams of air bubbles as they float to the surface. They also have the capacity to learn how to act on simple instructions, hence their role as performers in circuses. Sea lions are also used to run errands between underwater laboratories and the surface. They have been taught to carry tools to divers and to recover objects lost on the seabed by fixing a line to them. Recently, some sea lions have been trained to film whales underwater.

SEALS

SEALS, SEA LIONS AND WALRUSES are found in coastal regions throughout the world, including the polar seas, with a few species living in the open ocean or in freshwater lakes. Their ancestors were land-living mammals and they retain a link with their past by coming ashore to breed. However, in all other ways they are perfectly adapted to their aquatic lifestyle. Seals, sea lions and walruses are capable of prodigious feats of diving as they search for fish, crustaceans or squid.

Members of this group vary considerably in size. The ringed seal, *Phoca hispida*, weighs 200 pounds (90 kg), whereas the male southern elephant seal, *Mirounga angustirostris*, weighs about 3½ tons (3,600 kg).

Classification

Together, seals, sea lions and walruses constitute the order Pinnipedia. All the members of this order are characterized

Harp seals are born with a dense white coat of fur that darkens as they mature. Whereas the young of most seals are called pups, immature harp seals are known as whelps.

by sleek, streamlined bodies for efficient movement through the water and four modified limbs that act as flippers. The order is split into three families: the Phocidae, or true seals; the Otariidae, consisting of the sea lions and fur seals; and the Odobenidae, of which the walrus, *Odobenus rosmarus*, is the only living representative. The families Otariidae and Odobenidae have many features in common and are sometimes grouped together to form a superfamily: the Otarioidea. The members of this group diverged from a common ancestor that lived about 23 million years ago, and evolved quite separately from the true seals.

The Otariidae family contains five species of sea lions and nine species of fur seals, known collectively as the eared seals because of their visible ear flaps, a feature that distinguishes them from the Phocidae. The group includes the California sea lion, *Zalophus californianus*, a species often seen in zoos and marine aquaria. Sea lions and fur seals are anatomically similar, both having hind limbs that are jointed at the ankle, giving them more mobility on land than the true seals. Fur seals can be distinguished from sea lions by the thick coat of underfur that gives them their name.

CLASSIFICATION	
CLASS Mammalia	
ORDER Pinnipedia	
FAMILY Phocidae: true seals; Odobenidae: walrus; Otariidae: fur seals and sea lions	
NUMBER OF SPECIES 34	

Walruses are now the sole members of the once-varied Odobenidae family. Although they are actually more closely related to the otariids, they have a number of features in common with the phocids too, most notably the lack of visible ear flaps. However, with huge, curved tusks, bushy whiskers and a massive wrinkled body, it is impossible to confuse walruses with any other seals.

The Phocidae family is distinct from the other two families in the order Pinnipedia and evolved from a different ancestor, a weasel-like carnivore that lived about 25 million years ago. The group contains 19 species, including the southern elephant seal and the rare Hawaiian monk seal, *Monachus schauinslandi*. Members of the family Phocidae are sometimes known as earless seals because of their lack of external ear flaps. The phocids have hind limbs that are bent backward to form a tail fin, making them highly efficient swimmers, but they are slow and cumbersome out of water.

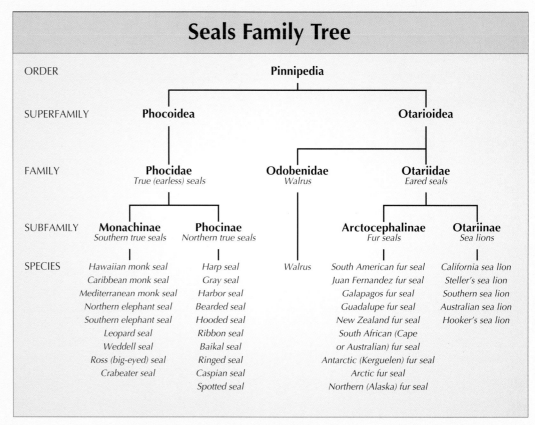

Seals Family Tree

ORDER		Pinnipedia			
SUPERFAMILY	**Phocoidea**			**Otarioidea**	
FAMILY	**Phocidae** *True (earless) seals*		**Odobenidae** *Walrus*	**Otariidae** *Eared seals*	
SUBFAMILY	**Monachinae** *Southern true seals*	**Phocinae** *Northern true seals*		**Arctocephalinae** *Fur seals*	**Otariinae** *Sea lions*
SPECIES	*Hawaiian monk seal* *Caribbean monk seal* *Mediterranean monk seal* *Northern elephant seal* *Southern elephant seal* *Leopard seal* *Weddell seal* *Ross (big-eyed) seal* *Crabeater seal*	*Harp seal* *Gray seal* *Harbor seal* *Bearded seal* *Hooded seal* *Ribbon seal* *Baikal seal* *Ringed seal* *Caspian seal* *Spotted seal*	*Walrus*	*South American fur seal* *Juan Fernandez fur seal* *Galapagos fur seal* *Guadalupe fur seal* *New Zealand fur seal* *South African (Cape* *or Australian) fur seal* *Antarctic (Kerguelen) fur seal* *Arctic fur seal* *Northern (Alaska) fur seal*	*California sea lion* *Steller's sea lion* *Southern sea lion* *Australian sea lion* *Hooker's sea lion*

The Phocidae family is further split into northern and southern phocids. Northern phocids are largely restricted to the Northern Hemisphere, and include Arctic species such as the bearded seal, *Erignathus barbatus*, and the ringed seal. Most northern phocids are 52–70 inches (1.3–1.8 m) in length, although bearded seals can reach 8¾ feet (2.7 m). Southern phocids are rather larger, commonly measuring 86–128 inches (2.2–3 m) in length, and are, with a few exceptions, restricted to the Southern Hemisphere. The group includes the world's most common pinniped, the crabeater seal, *Lobodon carcinophagus*, and the Caribbean monk seal, *Monachus tropicalis*, which is probably now extinct.

Senses

Seals, sea lions and walruses rely on their excellent sensory systems to locate their prey.

A seal uses its whiskers to sense vibrations underwater and in social interaction. For example, when the nasal whiskers point forward, they generally indicate aggression.

Eared seals probably favor shallower waters than true seals. However, the Australian sea lion (above) has been caught in traps at depths of 387 feet (118 m).

All species are heavily reliant on their vision, which is well developed. The large, near-spherical eyeballs optimize light intake, allowing the animals to see superbly in underwater darkness. Vision is also important on land, where seals are at their most vulnerable and rely on their keen senses to avoid predators. But the light is much brighter out of the water, and seals have to contract their pupils to the size of a pinhole in order to protect their eyes from damage.

Most seals and sea lions also have an excellent sense of hearing. Except for the absence of external ear flaps in walruses and true seals, the structure of the ear is similar to most other mammals. However, the bones of the middle ear are much larger than those of similar-sized land mammals, allowing the seal to precisely pinpoint the direction of sound. This not only enables seals to locate their prey, it also allows them to find each other and to communicate over long distances. Water is a much better conductor of sound than air, and seals can communicate with a series of grunts that can be heard up to 18 miles (30 km) away.

Walruses lack the keen hearing of their relatives, but they seem to compensate with a simple echolocation system. They emit a number of clicks and grunts, and appear to be able to work out the direction of their prey by the nature of the echo. Walruses also share with many other seals the ability to use their sensitive whiskers to find prey. At the base of the whiskers are nerve endings that can detect tiny vibrations passing through the water.

Unusually for carnivores, the seals, sea lions and walruses do not seem heavily reliant on their olfactory system (sense of smell). Nevertheless, scent may play an important role in mating rituals, since both males and females emit strong odors during the breeding season. Females also seem to identify their pups by scent.

Breeding

Nearly all types of seals and sea lions have an annual breeding cycle. Most species produce a single pup, or occasionally two, in spring or summer, although the precise timing varies considerably across species. At the start of the breeding season, females return to shore, seeking out a cove or a hole in the ice in the case of true seals, a sandy beach in the case of sea lions, and a rocky crag in the case of fur seals. In many species, males compete to establish hierarchy, territorial boundaries and access to females, performing noisy displays and sometimes engaging in fights.

Several days after coming ashore, a female gives birth and starts suckling her pup. Between 1 and 7 weeks after giving birth, she comes into estrus (heat) and mates with a dominant male. This allows mating to occur at a similar time to giving birth, minimizing the need for mass congregations on land, where seals are most vulnerable to predation. After mating there is a delay of about 3 months before the fertilized egg implants, giving the female time to recover from the strain of pregnancy and lactation (producing milk). Once the blastocyst (immature embryo) has implanted, the gestation period is 7–8 months, completing the annual cycle. Females suckle their pups for up to 1 year, depending on the species, although the pups become increasingly independent over this period.

The only exceptions to this annual breeding cycle are the walrus and the Australian sea lion, *Neophoca cinerea*, which have gestation periods of 18 months and 15 months respectively. This means that they can breed only once every 2 years.

Diving ability

All seals are able to stay underwater for long periods, and some can dive to great depths. Weddell seals, *Leptonychotes weddelli*, can remain submerged for 73 minutes and reach depths of 2,000 feet (600 m). Southern elephant seals may be the champion divers, with records of dives lasting up to 2 hours and reaching depths of 4,000 feet (1,200 m). Even an

Northern fur seal bulls have large harems, averaging 30 cows or more. A bull may occasionally force a cow into his harem but cannot effectively prevent cows from moving elsewhere.

average dive for these species may last 20–30 minutes. Sea lions and fur seals cannot remain underwater for as long, but they are nevertheless capable of staying below the surface for up to 10 minutes.

Seals have a number of adaptations that allow them to perform such feats. When diving they close their nostrils, and the pressure of the water then keeps them shut. By pressing the tongue and soft palate together, seals also close off the buccal (mouth) cavity, allowing them to catch food in their mouth without drawing water into their lungs.

Seals, sea lions and walruses can stay underwater for so long because they have the ability to build up stores of oxygen in the body. Their blood contains about three-and-a-half times as many erythrocytes (oxygen-carrying red blood cells) as that of humans. In addition, they have a high ratio of blood to body size. Twelve percent of a seal's body weight is made up of blood, compared to just seven percent for a human.

Most seals live in circumpolar regions, and consequently diving often takes place in very cold water. The animals have therefore also adapted to avoid excessive heat loss. Their bodies are streamlined, with appendages kept to a minimum, reducing the surface area that is exposed to the water. Seals are also well insulated. Their skin is covered by thick, wiry hair that traps an insulating layer of air. Beneath the skin is a

Walruses are gregarious, often basking together in large numbers. The size of the tusks places each walrus within a social hierarchy.

layer of fatty tissue known as blubber. Unlike fur, this retains its insulating properties at high pressure, allowing seals to stay warm even when diving to great depths.

Conservation

Throughout the 19th century and the early 20th century, seals, sea lions and walruses all suffered from heavy commercial exploitation. Traders valued fur seals for their thick, luxurious coats of underfur, while they hunted many other seal species for their hide, their meat and their blubber, which is a valuable source of oil. It was not until the 1960s, when television footage of seal hunts led to public outrage, that governments took serious action to restrict the slaughter. By that time some species had been brought to the brink of extinction, and one species, the Caribbean monk seal, had disappeared.

Hunting of seals, sea lions and walruses is now carefully regulated, and although commercial exploitation continues in some countries, it is no longer the threat that it was. As a result, some seal populations have recovered well over the last few decades, though human activities remain a threat to many species. Pollution from chemicals such as PCBs, DDT and heavy metals can weaken the immune systems of seals, making them vulnerable to disease. Purse-seine fishing and trawling often catch and drown the animals, while other species are culled to protect fish stocks. In addition, overfishing can have a devastating effect on the marine ecosystem, starving those predators that depend on fish for food.

The Mediterranean monk seal, *Monachus monachus*, is now listed as critically endangered by the I.U.C.N. (World Conservation Union). This classification came after an outbreak of disease, probably triggered by pollution, reduced the species' population to just a few hundred. The Hawaiian monk seal is listed as endangered, as is Steller's sea lion, *Eumetopias jubatus*, and the Baltic Sea population of the gray seal, *Halichoerus grypus*. A further 10 species or subspecies of seals are considered to be vulnerable, indicating that the world's populations of seals, sea lions and walruses are far from secure.

For particular species see:
- CRABEATER SEAL • ELEPHANT SEAL • FUR SEAL
- GRAY SEAL • HARBOR SEAL • HARP SEAL
- HOODED SEAL • LEOPARD SEAL • MONK SEAL
- RINGED SEAL • ROSS SEAL • SEA LION
- WALRUS • WEDDELL SEAL

SEA MOTH

The longtail sea moth is the largest sea moth. As in other species, the bony plates that cover its body make effective defensive armor.

There are five species of sea moths belonging to two different genera. Four species range from the coast of South Africa, around the shores of the Indian Ocean to eastern Australia, Indonesia and Micronesia, and as far north as China and Japan. The Hawaiian sea moth, *Eurypegasus papilio,* occurs around Hawaii. Sea moths are entirely absent from the Atlantic. However, studies may show that all species have a wider distribution than is known at present.

One of the best-known species is the longtail sea moth, also known as the batfish, sea dragon or dragonfish. These names are also variously applied to the other species. The longtail sea moth ranges from South Africa east to Japan.

Flightless moths

Despite their name, there seems to be no question of sea moths flying because the fin muscles are not strong enough for flapping flight and the body is too heavy for gliding. It is possible that these large fins are used to confuse a possible predator when spread out suddenly, giving the impression that the sea moth is a much larger fish. The pelvic fins are long and hook-shaped.

Lagoons of seagrass

Sea moths live in shallow coastal waters down to depths of about 20 feet (6 m). They prefer sheltered lagoons, estuaries and bays with beds of seagrass or seaweed and sandy or silty bottoms. Sea moths do not appear to be strong swimmers and probably do not travel far. Some scientists have suggested that these fish can move over the sea bottom using their pelvic fins as legs and their large pectoral fins either in a swimming action or to give stability to the body. Sea moths are nevertheless widely dispersed across the oceans, but this may be because their larvae are carried thousands of miles by ocean currents, as part of the plankton.

Bottom feeders

The mouth is situated on the underside of a long snout, which indicates that sea moths are bottom feeders. The small size of the mouth means they are restricted to extremely small invertebrates. Sea moths feed from the surface of the seabed, taking minuscule animals that live among the sand grains or crawl across the top of the substrate.

WHEN THE FIRST MARINERS set out from western Europe to explore the waters of East Asia and the South Pacific, they brought back the dried bodies of sea moths as curios and used them as decorations and ornaments. The sea moth looks a lot like the pipefish and the seahorses (both discussed elsewhere in the encyclopedia), except that it has large, winglike pectoral fins. However, it is entirely unrelated to these other fish.

Sea moths are usually about 5 inches (13 cm) long, but the largest, the longtail sea moth, *Pegasus volitans,* may be up to 7 inches (18 cm). The squat body is pale brown with white spots and greenish blue fins. It is covered with bony plates arranged in concentric rings. The toothless mouth is small and instead of being at the tip of the elongated snout, as in the seahorse, is farther back underneath the snout. On each side a narrow gill opening is situated in front of the pectoral fin, and the gill cover is formed of a single plate, within which there are four gills. There is no swim bladder. The dorsal and anal fins are short and soft-rayed, but the pectoral fins are very large and winglike, suggesting some sort of relationship with the flying gurnards and giving rise to the name of sea moth.

DRAGON SEA MOTH

CLASS	**Osteichthyes**
ORDER	**Gasterosteiformes**
FAMILY	**Pegasidae**
GENUS AND SPECIES	***Eurypegasus draconis***

ALTERNATIVE NAME
Short dragonfish

LENGTH
Up to 4 in. (10 cm)

DISTINCTIVE FEATURES
Broad, depressed body contained in bony plates; tail free for some movement; small, toothless mouth, protrusible jaws; lobelike gill filaments with tufts; no swim bladder

DIET
Mainly small crustaceans and worms

BREEDING
No details known

LIFE SPAN
Not known

HABITAT
Lagoons, often among algal or seagrass beds; also found on sand or silt bottoms, frequently in bays or estuaries

DISTRIBUTION
Coastal regions of Indian and southeastern Pacific Oceans

STATUS
Generally common

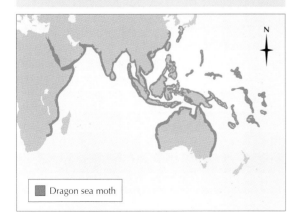

Dragon sea moth

The larvae of sea moths are rarely found, but they are occasionally taken off the coast of India in tow-nets used for collecting plankton. The larvae begin to develop their idiosyncratic morphology, their winglike pectoral fins and hooklike pelvic fins, at a very early stage, when they are only 1 millimeter long. At this stage though, there is no snout, the upper jaw is tilted slightly upward and the lower jaw projects beyond it, which suggests that the larvae feed on small plankton.

Prisoners of the Indian Ocean

In the distribution of the sea moth can be seen a good example of how a species of marine animal is limited by physical barriers. The sea moths have dispersed themselves throughout the coastal waters of the southern Pacific and Indian Oceans, all the way west to South Africa. So why have they not reached the Atlantic?

The answer almost certainly is that they could not get past the cold barrier in the sea off the Cape of Good Hope. At this point, the warm Agulhas Current meets the cold Benguela Current. The fauna on either side of the junction of these two currents are quite different. The fauna in the area of the Benguela Current has more in common with that of the Mediterranean Sea, while that in the area of the Agulhas Current is of the same kind found in the Indian Ocean. It is almost as if there existed in the sea an actual impenetrable barrier. Around Japan, at the other extreme of sea moth distribution, there is a similar pair of cold and warm currents, less sharply defined, but still offering an obstacle to the spread of warm-water species.

Variously called the short dragonfish or dragon sea moth, this sea moth slowly scans the seabed for small crustaceans, worms and other inhabitants of the sand.

SEA MOUSE

ESPITE ITS COMMON NAME, the sea mouse is not a rodent, although it resembles a mouse and moves over the sand with a mouselike creeping movement. Its back also has a similar coloration to that of a mouse. However, the animal is actually a segmented worm, one of the marine bristleworms related to the ragworms (discussed elsewhere).

The sea mouse of the coasts of Europe grows up to about 8 inches (20 cm) long and about 3 inches (7.5 cm) across at its widest point. It is flattened from above downward and is more or less oval in outline. Its back is covered with 15 pairs of large scales, but these are hidden from view by a feltwork of densely packed bristles. These bristles are gray along the middle of the back but along the margins they are iridescent. There are two short tentacles in front and a pair of stalked eyes.

The undersurface shows the rings that are characteristic of earthworms and marine bristleworms, and on each side of each segment is a group of stout bristles. In effect, therefore, a sea mouse is a ragworm with the body broadened and thickened, covered with large overlapping scales and further coated with a layer of bristles. Intermediate between the ragworm and the sea mouse are scale worms, which are broader than the ragworm but not as broad as the sea mouse.

The large scales protecting the back of the sea mouse are invisible beneath its thick covering of bristles.

Scale worms have an armor of scales covering the back but lacking the felting of bristles. They are found in shallow seas or on the shore.

Sand burrows

Sea mice live in shallow seas just below low tidemark and are sometimes brought up in a dredge or in a fisher's trawl. They are most often seen after storms, when they are occasionally thrown up on the sandy shore and left stranded, dead or dying, by the thousands by the ebb tide. A storm must be violent to do this, however, because normally the sea mouse burrows in the sand, pushing itself along with the tufts of strong bristles at each side, leaving only its tail exposed at the surface.

When it is thrown up on the shore by the action of waves, or when it is first dug out of its burrow, the sea mouse's back is sandy or muddy. After it has been placed in an aquarium for a while, the grains of sand or mud are washed off, revealing its true colors. Away from the gray back, the bristles along the edges glisten with metallic colors, gold or red, yellow, orange, scarlet or crimson, lilac or ultramarine, according to which way the light strikes them. In the sea mouse's submerged burrow these colors are never apparent.

The feltwork of bristles and the scales are most important to this burrowing worm. While it is in its burrow the sea mouse, with its tail end at the surface of the sand or mud, draws in a current of water by muscular movements of its undersurface. This water passes onto its back, where its gills lie protected by the scales from the weight of sand. The felt of bristles over the scales keeps out sand grains that would otherwise clog the gills, particularly when, as happens periodically, the scales are drawn down to squeeze out the water from which oxygen has been extracted.

Feeding and breeding

The sea mouse burrows to obtain carrion (the dead and decaying bodies of small animals or parts of larger animals) that have become submerged in the sand. As it moves slowly forward, plowing through the sand or mud, the sea mouse swallows pieces of dead flesh. Just behind the mouth is a muscular tube that squeezes and grinds the food, breaking it up into small

SEA MOUSE

PHYLUM	**Annelida**
CLASS	**Polychaeta**
ORDER	**Phyllodocida**
FAMILY	**Aphroditidae**
GENUS AND SPECIES	***Aphrodita aculeata***

LENGTH
Up to 8 in. (20 cm)

DISTINCTIVE FEATURES
Short, wide, flattened bristleworm; underside divided into wormlike segments; upper surface has covering of bristles, long and coarse along sides, feltlike on back

DIET
Scavenges dead and dying animals

BREEDING
Sexes separate; breeding season: fall and winter. Larva has never been seen; young may hatch directly from egg; alternatively, may settle after a few weeks in plankton.

LIFE SPAN
Probably up to 3 years

HABITAT
Sandy seabeds; sometimes in lower intertidal zone but usually up to 330 ft. (100 m)

DISTRIBUTION
Coastal waters of Europe and northern Africa; Black Sea

STATUS
Locally common

Sea mouse

The underside of a sea mouse clearly shows the segmentation that is characteristic of marine bristleworms and terrestrial earthworms.

history. Although scientists have been studying it for over a century, the sea mouse has been seen laying eggs on very few occasions. Its larvae have not been found, so scientists have to assume the sea mouse has either a very brief free-swimming life or that the young sea mouse hatches directly from the egg.

Self-defense

The sea mouse is probably eaten by larger animals such as cod and some of the bottom-feeding fish, particularly when the sea mouse is small. A large sea mouse with its many bristles is probably not a very attractive prospect as a meal if its near-relative, the fireworm *Hermodice carunculata*, which lives on the coral reefs in the Caribbean, is any indication. The fireworm has fine white bristles that are very brittle and break easily into short lengths. An animal trying to eat it risks having the inside of its mouth and stomach lined with a substance similar to glass pins.

People who have come into contact with the fireworm describe how merely brushing the skin against it gives a burning sensation that lasts as long as a real burn. Rubbing the wound makes matters worse, as this action drives the bristles further into the skin. This means of defense must be effective because it is found elsewhere in the animal kingdom, notably in the hairy caterpillars of certain moths and butterflies. The sea mouse is not normally very difficult to handle, but nevertheless must be unpleasant for predators. It is occasionally found in the stomachs of a variety of bottom-dwelling fish, but not as commonly as its abundance would suggest.

particles. Behind the muscular tube the true gut has many fine tubes running off sideways, into which the food passes to be digested.

Apart from its shape and its way of breathing and feeding, the sea mouse resembles other bristleworms. However, there is still considerable scientific debate about its life

SEA OTTER

Using its chest as a dining table, the sea otter skillfully uses rocks as tools to break open its shellfish prey. In this case, the victim is an abalone.

THE SEA OTTER WAS ONCE hunted in American waters to the brink of extinction for its valuable fur. It is an exclusively marine animal whose head and body measure up to 47 inches (1.2 m) long. It has a 10–14½-inch (25–37-cm) tail and weighs up to 99 pounds (45 kg). It has one of the densest furs of any mammal. The fur is thick and glossy, varying in color from rufous to dark brown and sprinkled with white-tipped hairs. The head, throat and chest are creamy gray. The sea otter has a large, blunt head, a short, thick neck with short, pointed ears almost hidden in the fur, and small eyes. Its hind feet are long, broad, webbed and flipperlike, whereas its forefeet are comparatively small and have retractile claws.

The sea otter is the only carnivore with four incisor teeth in the lower jaw. Its molars are broad, flat and well adapted for crushing the shells of crabs, sea urchins and other shellfish on which it feeds. Unlike most marine animals, a sea otter has no layer of fat or blubber under the skin to keep it warm. When resting or sleeping, the sea otter would lose too much body heat but for the air trapped in its fur. This insulating layer is so essential that if the fur is damaged, the sea otter may die from cold and exposure.

Hunted ruthlessly

The sea otter once ranged along the Pacific coasts in vast numbers, from the Kurile Islands, along the coast of the Kamchatka Peninsula and Alaska, among the Commander and Aleutian Islands and down the western coast of North America to lower California. In the 18th century they were ruthlessly exploited by fur hunters, and thousands were killed for their valuable fur until very few were left. By the beginning of the 20th century the sea otter had become so rare that its pelts were regarded as the most valuable of all furs, fetching as much as $1,000.

In 1910 the United States government introduced a law prohibiting the capture of sea otters within U.S. waters, and by 1911 other interested governments had followed suit. Today, the sea otter has been increasing by 5 percent per year and has returned to many of its colonies from the Kuriles to southern Alaska. However, there are no sea otters on the western coast of North America except around Monterey, California.

Feeds in kelp forests

The sea otter spends its time, usually in small groups, in shallow waters off the rocky mainland or island shores, particularly favoring the waters

SEA OTTER

CLASS	**Mammalia**
ORDER	**Carnivora**
FAMILY	**Mustelidae**
GENUS AND SPECIES	**_Enhydra lutris_**

WEIGHT
33–99 lb. (15–45 kg)

LENGTH
**Head and body: 39–47 in. (1–1.2 m);
tail: 10–14½ in. (25–37 cm)**

DISTINCTIVE FEATURES
**Stocky body; large, blunt head; short, thick
neck; short, pointed ears; small eyes;
webbed hind feet; very dense, glossy fur;
mainly rufous to dark brown in color, with
creamy-gray head, throat and chest**

DIET
**Mainly shelled invertebrates such as
abalones, sea urchins, clams, mussels and
crabs; occasionally small fish and octopuses**

BREEDING
**Age at first breeding: 5–6 years (male), 3–4
years (female); breeding season: all year,
peaking January–May and early autumn;
number of young: 1 to 3, but only 1
survives; gestation period: 120–160 days;
breeding interval: usually 2 years**

LIFE SPAN
Up to 23 years

HABITAT
**Rocky coasts, usually within ½ mile (800 m)
of shoreline; strongly associated with kelp
forests of California**

DISTRIBUTION
Eastern coasts of Japan, Canada and U.S.

STATUS
**Endangered; estimated population: fewer
than 150,000**

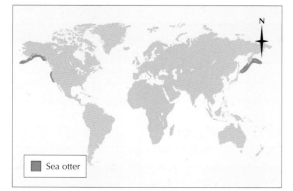

Sea otter

around kelp beds. Although it only occasionally
ventures more than 1 mile (1.6 km) from land, it
rarely comes ashore, except in violent storms.
The sea otter is active in the day, feeding,
playing, swimming or floating on its back. It
swims belly-down only when in a hurry or to
avoid a predator. When resting, the sea otter
wraps strands of kelp around its body to serve as
anchors and stop it drifting off in its asleep.
Sometimes it sleeps with its hands over its eyes
as if to shade them from the moonlight.

Using tools

The sea otter feeds mainly in the early morning
and evening on sea urchins, clams, crabs,
mussels, abalones and other shellfish. Fish and
octopuses are occasionally taken. It dives for its
food, sometimes to depths of as much as 160 feet
(48 m), and floats on its back while eating, using
its chest as a table. It is one of the few mammals
to make deliberate use of a tool; when diving for
food, it will sometimes bring up a flat stone and,
placing this on its chest, use it as an anvil to
smash the shells of mussels and other shellfish. It
holds up the mollusk between its paws and
crashes it down repeatedly onto the stone until
the shell breaks. Large crabs are eaten piecemeal,
their legs torn off one by one, and are devoured
while the crab runs about on the otter's chest. Sea
otters require a lot of food. They eat around a
quarter of their body weight each day: in some
individuals as much as 20 pounds (9 kg).

*The sea otter remains
an endangered species
although its numbers
are recovering in
some areas.*

Slow breeders

Mating, which takes place throughout the year and in the water, is preceded by an elaborate courtship. There is delayed implantation, so the true gestation period is 120–160 days, after which one to three pups are born, of which only one survives. The pup is born on shore in an advanced stage of development, well furred, with its eyes open and with a complete set of milk teeth. The mother immediately carries her pup into the water and gives it constant and careful attention, nursing and grooming it on her chest as she swims or floats on her back. The pup does not leave her until it is about 8 months old, probably relying on her to teach it to hunt. The reproductive rate of sea otters is slow because males do not breed until they are 5–6 years old, whereas females are 3–4 years old before they breed, and they bear pups only every other year.

Humans were its worst enemy

The sea otter has formidable natural predators in the killer whale and various sharks, but in the past its worst predator was humans, who hunted it for its fur. The Native Americans made little difference to the sea otter's numbers. It was a different story once the animal was exploited for the fur trade by European settlers.

Holding without hands

Very few animals, only two dozen or so, use tools, yet many could do so. Two things are needed for this: the ability to hold the tool and the "idea" of using it. One of Darwin's finches, the woodpecker-finch, *Camarhynchus pallidus*, probes insects from holes using a cactus spine held in its bill. All birds have bills, many feed on insects and a number of them live where cacti grow. Yet the woodpecker-finch is the only one with the necessary behavior pattern that leads it to break off a spine and use it as a probe.

There are many mammals, apart from monkeys and apes, with handlike forefeet, and yet they use no tools. The toes on a raccoon's forefeet, for example, are very supple and finger-like, and a raccoon can even untie knots in string. Yet it does not use a tool. Some mammals will pick things up with their forefeet, the beaver being an example, but they do not use their feet to wield tools.

The paws of a sea otter are small and the toes are very short. In fact, the paws are little better than stumps at the ends of the forelegs. Nobody could guess from looking at them that the sea otter could manipulate a tool. Yet the sea otter uses a stone to crack open mollusk shells on its chest as efficiently as it possibly could with these stumps, because it has the inborn impulse (or innate behavior pattern) to do so.

It makes one wonder what the sea otter might have been able to do if it had long, slender fingers and opposable thumbs. Perhaps it would have been capable of nothing more than it does now, because its behavior pattern is limited to cracking open clams, mussels, sea urchins, crabs and other shellfish.

Sea otters are strong swimmers in the open sea, where they are sometimes threatened by oil pollution. Oil clogs their layer of insulating fur, rendering it useless.

SEA PEN

THE SEA PENS ARE RELATIVES of anemones and sea firs. Some resemble old-fashioned quill pens stuck in the mud on the sea floor, hence their common name. They vary widely in shape; some are flattened almost like a broad leaf at the top of a stalk and have earned the name of sea pansies. All sea pens have the same basic shape: a main central stem, from the upper part of which spring either polyps or branches bearing tiny polyps. In those species that most resemble a quill, the branches springing from each side of the stem are closely packed and resemble the vane of a feather. In some sea pens the central stem is very thick and the polyps are scattered on it randomly. In others it is long and slender, with a few polyps in a bunch at the tip or scattered in pairs well spaced out along its length. Each sea pen is made up of a primary polyp that forms a central stem, from which the secondary polyps bud. The mouth and tentacles of the primary polyp usually degenerate so this first polyp becomes no more than a fleshy stem carrying the other polyps. The fleshy stem is strengthened by a central horny rod.

The smallest sea pens are only a few inches high, whereas the largest are about 6½–10 feet (2–3 m) tall. They are mostly colored yellow, orange, red or brown. The colors are mainly due to the pigments of calcareous spicules (tiny rods made of calcium) scattered in the flesh and forming a skeleton. The shape and pattern of the spicules vary from one species to another.

Sea pens are found worldwide. In some areas there are species living between the tidemarks, but most are in fairly deep water and some live at depths of 16,400 feet (5,000 m) or more.

Peristaltic movement

Sea pens can move and give off light. Their nearest relatives are dead man's fingers (discussed elsewhere). Each polyp is separated by partitions dividing the cavity into four longitudinal canals. The central stem also has longitudinal canals. Through these, with the help of muscles in the stem, water can be taken in and given out, so the whole colony is able to swell or deflate. The muscles work by peristaltic movements, that is, by wavelike motions such as those passing along the wall of the intestine or along the body of an earthworm. By the combined use of hydrostatic pressure and peristaltic movements, sea pens can move far more than might be suspected from seeing them standing on the sea floor with the lower end of the "quill" buried in mud or sand as an anchor.

Retreating into mud

Sea pens living on mudflats, such as *Stylatula elongata* on the American Pacific Coast, are exposed when the tide goes out, but can pull down their upper part as the tide ebbs, pumping out water to make themselves smaller until they are buried in the mud. They can also pull down the central stem until only an inch or two is exposed above the surface of the mud. Should the stem be damaged or broken off, the sea pens lay down an extra piece at the other end, restoring the rod to its full length. As the tide comes in the central rod is pushed up again, and by drawing water into the long canals, the soft body is inflated and pushed up the central rod, back to its normal position. Sea pens living in deeper water have no need to carry out this maneuver. However, they often need to move from one place to another when food is scarce. By a combination of muscular movement and pumping water in and out, a sea pen can pull the lower part of its stem out of the mud or sand, move over the surface, push its stem in again and

Sea pens feed on small animals up to a few millimeters long. They catch them using stinging cells located in the polyps.

SEA PENS

PHYLUM	**Cnidaria**
CLASS	**Anthozoa**
ORDER	**Pennatulacea**
FAMILY	**Funiculinidae; others**
GENUS AND SPECIES	***Funiculini quadrangularis*; many others**

LENGTH
Up to 6½ ft. (2 m) or more

DISTINCTIVE FEATURES
Very long and slender; quadrangular stem

DIET
Tiny animals

BREEDING
Little known; separate sexes; fertilization in sea water; planulae (free-swimming, ciliated larvae) settle on sand or mud with rich organic film

LIFE SPAN
Not known

HABITAT
Muddy seabeds from 66 ft. (20 m) to several hundred feet

DISTRIBUTION
Worldwide, but not known in detail; North Atlantic, Pacific and Indian Oceans

STATUS
Locally common

Many sea pens (genus Pennatula, *above) are found on flat sand or muddy areas. As a consequence they are easily damaged on a wide scale by trawling.*

lift the rest of itself into its usual position. Sea pansies, genus *Renilla*, of the Atlantic, Gulf and south California coasts, can do the same. They are short, fat sea pens that live in shallow, turbulent, sandy areas. They can also bury themselves partially by pumping out water and pulling down with their muscles.

Free-swimming larvae

Sea pens feed on very small animals in much the same way as the dead man's fingers and other small polyps do. They also breed in a similar way to their relatives, by free-swimming larvae. These are oval in shape and covered with cilia, the orderly beating of which drives the larvae through the water. After a while the larvae settle on the seabed and change into polyps. The polyps grow rapidly in length and form the main stems from which other polyps are budded.

Underwater illuminations

Pennatula phosphorea is a sea pen living on muddy bottoms in fairly deep water in the North Atlantic. It is sometimes brought up in trawls. When it is touched by something it lights up at the point of contact, and the light spreads until the whole animal is glowing. *Funiculina quadrangularis* lives in the same area, and probably ranges worldwide. It is 6 feet (1.8 m) tall. When its stem is touched, a glowing light travels up and down the stem, like a neon sign. Some sea pens light up all over the body, whereas others light up only in certain parts. In some sea pens when the base is touched, light will travel from there to the other end, or if the top is touched, light travels down and fades out at the base. If the two ends are touched simultaneously, the light travels from both ends and meets in the

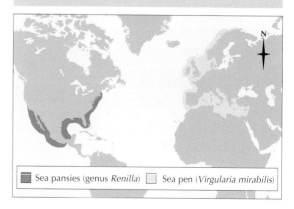

Sea pansies (genus *Renilla*)   Sea pen (*Virgularia mirabilis*)

middle, where it fades out. The strength of the light produced depends on the strength with which the sea pen is touched, causing it to give out mucus filled with luminous granules. Sea pens have no eyes to detect this light but it may be that the luminosity frightens away animals that might otherwise eat the sea pens. The color of the light varies from species to species and may be yellow, green, blue or violet.

SEA SLUG

SLUGS ARE SNAILS WITHOUT shells, or with very reduced shells. They have evolved separately, both on land and in the sea, to give rise to a number of unrelated groups. For this reason, the rich variety of sea slugs, including some of the most brightly colored of sea animals, does not constitute any single natural group of mollusks. Most sea slugs belong to one subclass of the gastropods, the opisthobranchs, along with the sea butterflies and sea hares, which have much-reduced shells. The opisthobranchs were formerly divided into two groups, one made up mainly, but not entirely, of forms with shells, called the tectibranchs, and the other of slugs, called the Nudibranchia. Nudibranch persists as a useful common name for this group of slugs. However, Nudibranchia as a proper name is now applied to a more restricted group, which this article describes.

As well as abandoning the shell typical of their ancestors, sea slugs lost the mantle cavity and contained gills that went with it. The ancestral twisting of the body disappeared too, and the mass of viscera formerly housed in the shell was incorporated in the foot. Near the front of the body are sensory tentacles, while outgrowths of various kinds often sprout from the surface elsewhere. These growths aid in respiration, but they may also have other functions. There are about 45 families of nudibranchs, and only a brief overview of their variety is possible here.

Camouflaged among their food

One of the most common sea slugs on the coast of Europe is the sea lemon, *Archidoris pseudoargus*, which has a tough yellow body mottled with patches of brown, red or green and is the size of half a lemon. It has a pair of sensory tentacles near the front and a circlet of nine retractile, feathery gills around the anus, which lies on the back toward the hind end. Hermaphroditic, like all other sea slugs, the sea lemon lays its eggs in frilled white coils of jelly 16 inches (40 cm) long, from which the planktonic larvae emerge.

The sea lemon is not always easy to see because it rests among the green or yellowish sponges, *Halichondria*, on which it feeds. Not many animals eat sponges, but a number of relatives of the sea lemon do, including *Rostanga rufescens*, which matches the brick red of another sponge, *Microciona*. The color of this sea slug may be due to pigments from the sponge it eats.

Many sea slugs, such as the rose nudibranch, Hopkinsia rosacea (below), of California, are the same color as the substrate on which they live and feed, thereby gaining a form of camouflage.

Another sea slug, *Archidoris flammea*, which is red, feeds only on the red sponge *Hymeniacidon sanguinea*. The sea slug *Calma glaucoides* is ½ inch (1.3 cm) long and silvery gray. Its color is very similar to that of the eggs of certain rock-pool fish and the sea slug feeds on the eggs, sucking the yolk from them. Other, more distant relatives of the sea lemon suck the juices of sea squirts and moss animals. Almost all nudibranchs are carnivorous, feeding on sponges, bryozoans, hydroids, soft corals, anemones, ascidians and barnacles.

Another European sea slug, growing to a length of 8 inches (20 cm), is *Tritonia hombergi*. It has a line of branched, feathery gills along each side of the body. However, they are not simply gills, for *Tritonia* feeds on the white variety of dead man's fingers (discussed elsewhere), and the gill tufts resemble the polyps of its prey, helping to conceal the sea slug from predators.

Swimming slugs

Although most nudibranchs move by means of waves of muscular activity in the sole of the foot, there are some that swim. *Dendronotus* does so by flexing its body from side to side. The foot of this hydroid eater is reduced to a narrow groove, and the upper side of the body bears many branched brown and red gills. *Scyllaea* swims in much the same way, but hangs upside down from the weight of its tentacles and its four long dorsal papillae. *Melibe* and *Tethys* swim in a different way, by means of rows of leaflike paddles arranged in a line down each side of the back, each paddle with a pair of gills at its base.

These two nudibranchs each have a cowl-like hood at the front of the body that moves from side to side to ensnare swimming crustaceans.

Other swimming sea slugs include *Phyllirhoe*, which is flat and has no foot or outgrowths other than sensory tentacles. It stays in the plankton, swimming along by graceful undulations and feeding on the Portuguese man-of-war. It is semi-transparent and dotted with tiny, light-producing organs. Its body may contain zooxanthellae (symbiotic, chiefly marine plankton that live within the cells of other organisms). A few sea slugs float freely near the surface of the sea, buoyed up by gas in branches of the gut. These feed on jellyfish such as *Velella* and *Porpita*. From *Velella* they acquire a bluish color, which fades if the sea slugs are deprived of their prey.

Found in the seas off Papua New Guinea, Chromodoris bullocki is named for its large, hornlike tentacles.

SEA LEMON

PHYLUM	**Mollusca**
CLASS	**Gastropoda**
ORDER	**Nudibranchia**
FAMILY	**Archidorididae**
GENUS AND SPECIES	***Archidoris pseudoargus***

LENGTH
Up to 4¾ in. (12 cm)

DISTINCTIVE FEATURES
Body covered in tubercles (prominent bumps); yellow, green, brown and red patches; up to 9 gills, each with 3 main branches and numerous finer brown pinnules (featherlike appendages)

DIET
Sponges, especially *Halichondria panicea*

BREEDING
Hermaphroditic (possessing both male and female sex organs), but dependent on cross-fertilization; eggs laid in long coiled ribbons; hatch into veligers (ciliated larvae); larvae settle onto edible substrate and develop

LIFE SPAN
Up to 1 year

HABITAT
Commonly lower intertidal zone down to 660–990 ft. (200–300 m), usually on sponges

DISTRIBUTION
European coastlines

STATUS
Common

Secondhand weapons

Rock pools often contain an organism that has the appearance of an elongated sea anemone but that is actually a sea slug with its back covered by tentacle-like papillae. In fact, these slugs feed on sea anemones and make use of their prey's undigested stinging cells for their own defense. Having been eaten by the sea slugs, the stinging cells pass unharmed through the stomach up along branches of the digestive system and into the dorsal papillae. Eventually the stinging cells collect in sacs, each opening by a pore at the tip of an appendage.

The slug cannot discharge the stinging cells at will, as an anemone can; these cells only come into action when the papillae are forcibly detached. Armed in this way, a sea slug becomes a much less attractive prospect to predators, and the conspicuous yellows, reds and pinks of some species may also serve as warning colors.

However, it is possible that the coloring may serve only for concealment. The light grayish brown of *Aeolidia papillosa*, for example, which grows to a length of about 3 inches (7.5 cm), blends with that of its prey, the opelet or snake locks anemone, *Anemonia sulcata*, although this may be greenish in color.

Sea slug slime

It seems that stinging cells become covered with the sea slug's slime, and in this way pass through the stomach without harming the organism. It seems that this slime is also used when the slug is feeding on a sea anemone or other nettlesome animal. In tests, scientists found that when they dropped a sea slug onto the extended tentacles of an anemone it remained unharmed, even though the anemone discharged its stinging cells at it. When a sea slug starts to feed on the anemone, the latter may bend its tentacles over to the intruder, but the sea slug simply covers them with slime and continues to feed.

Labor-saving feeding

Sea slugs appear to utilize their food to the utmost. There are, for example, several species of sea slugs that feed on small, delicate green seaweeds. These seaweeds contain chloroplastids, small bodies that hold the chlorophyll by which the plants manufacture their food. These chloroplastids are not digested by the sea slugs but pass into diverticula, or blind tubes, leading off from the gut, where they continue to photosynthesize. By feeding radioactive tracers to sea slugs, scientists have shown that the carbohydrates that the chloroplastids manufacture pass into the cells of the sea slug and serve as food for the latter.

This partnership between sea slugs and chloroplastids appears to be simply a labor-saving device on the part of the sea slugs. They are surrounded by their seaweed food and do not have to rely on their relationship with the chloroplastids to obtain nourishment.

Native to Egyptian waters, **Chromodoris quadricolor** *is one of the most colorful sea slugs. Sensory tentacles grow near the front, while other outgrowths help in respiration.*

Index

Page numbers in *italics* refer to picture captions.
Index entries in **bold** refer to guidepost or biome and habitat articles.

Page numbers in *italics* refer to picture captions. Index entries in **bold** refer to guidepost or biome and habitat articles.

Page numbers in *italics* refer to picture captions. Index entries in **bold** refer to guidepost or biome and habitat articles.